CURIOUS ENCOUNTERS
WITH THE NATURAL WORLD

CURIOUS ENCOUNTERS
WITH THE NATURAL WORLD

FROM GRUMPY SPIDERS TO HIDDEN TIGERS

MICHAEL R. JEFFORDS AND SUSAN L. POST

Foreword by Peter H. Raven

UNIVERSITY OF ILLINOIS PRESS
Urbana, Chicago, and Springfield

All photographs by the authors unless otherwise attributed.
Partial funding provided by A Prairie Gallery, Champaign, IL.

Library of Congress Cataloging-in-Publication Data
Names: Jeffords, Michael R., author. | Post, Susan L., author.
Title: Curious encounters with the natural world : from grumpy spiders to hidden tigers / Michael R. Jeffords and
    Susan L. Post ; foreword by Peter H. Raven.
Description: Urbana : University of Illinois Press, [2017] | Includes bibliographical references and index.
Identifiers: LCCN 2016054029 (print) | LCCN 2016055685 (ebook) | ISBN 9780252082665 (pbk. : alk. paper) |
    ISBN 9780252099670 (e-book)
Subjects: LCSH: Natural history—Miscellanea. | Curiosities and wonders.
Classification: LCC QH45.5 J44 2017 (print) | LCC QH45.5 (ebook) | DDC 508—dc23
    LC record available at https://lccn.loc.gov/2016054029

WITHOUT NOTES,
LIFE IS FLEETING.

—Luann Wiedenmann, journal entry, Shamvura Camp, Namibia, June 29, 2008

CONTENTS

XIV

FOREWORD
Peter H. Raven

XVIII

ACKNOWLEDGMENTS

1

INTRODUCTION:
NATURALISTS AFIELD—A LIFETIME OF CURIOSITY

14

CHAPTER 1. AMASSED NATURE

46

CHAPTER 2. ENVIRONMENTAL FLUIDS

58

CHAPTER 3. HIDDEN WORLDS

92

CHAPTER 4. MOMENTS IN TIME

120

CHAPTER 5. PHYSICAL PHENOMENA

136

CHAPTER 6. SELDOM WITNESSED

170

CHAPTER 7. TRULY BIZARRE

194

CHAPTER 8. JUST BE CURIOUS

212

CHAPTER 9. WHO KNEW?

244

CHAPTER 10. BE PREPARED

258

CONCLUSION: THE END OF THE LANE

262

EPILOGUE: UNEXPECTED SURPRISES

266

GLOSSARY

272

REFERENCES

276

INDEX

FOREWORD

When the world began to open to explorers and adventurers about five centuries ago, these voyagers were astonished by the extraordinary biological diversity they encountered and all the wonderful traits that it exhibited to the curious eye. When our ancestors remained for the most part in their "home ranges," they would certainly still have delighted with what they saw. With the increasing urbanization of the global population, people tend to narrow their sights and to believe that whatever they need must be located near them. That unfortunate definition of "what they need" deprives them of what is surely one of life's greatest pleasures, that of experiencing the endless, mysterious, and often gloriously beautiful workings of nature in place.

Michael Jeffords and Susan Post have here provided a rich, lovely, and readable account of many of their own adventures, and those adventures make delightful reading for anyone with even a small amount of curiosity about the true nature of the world in which we live. We inhabit a unique planet that supports our lives and the lives of millions of other kinds of organisms, very poorly known or understood by us, that make it possible in their aggregate and complex activities and interactions for us to live here on Earth.

Early in their lives, Mike and Sue realized the joy of collecting, of recognizing what was out there, seeing how it behaved, and always searching for the unique missing species or bit of behavior that amazed and pleased their youthful curiosity. Having started at a comparable age—around six years old—collecting insects myself and being constantly amazed by what I could find, I relate to their tales ever so easily. It is an incredible deprivation, as Richard Louv has pointed out so clearly, for children to miss this kind of pleasure. Their life histories and the diversity of their form, their habits, their abundance, and the places they each lived gave me insight into the diverse array of beetles that I came to know in central California, and there was always something new for me to see. I could easily understand Mike's growing reluctance to have to kill insects to "have" them and began to look for other ways to relate to them myself.

Like many women, young Sue experienced limitations as to what she was "supposed" to do, as compared with the wide range of activities open to men. Insect collecting was not ruled out for girls, however, and she was soon preoccupied with gathering interesting specimens, too. While she and Mike were students at the University of Illinois (UI), an intellectual paradise, their common interests and major drew them together. Just as I did myself, Sue drifted into the study of plants, in her case spurred on by photography. The pleasures that they shared in exploring the world stand out vividly in the pages of this volume.

Both Sue and Mike are photographers of unusual talent, but this book has to do more with collecting "curiosities" than with the joys of photography. Nonetheless, their essays are illustrated with beautiful examples of their art, photographs and sketches that bring us closer to the objects of the short essays that fill the pages of the book's chapters. What they have succeeded in creating is a museum, populated not with examples of the taxidermist's and exhibit designer's art but with verbal gems that they have created, depicting the interactions of life in the world at large—remarkable images of its inhabitants and the ways they and their activities merge to form a living whole. Each chapter consists of a series of essays that they have crafted to highlight their observations of what they found going on in that particular place. Many of us would not have noticed the coevolutionary interactions that they photograph and sketch so well, but seeing them vicariously with outstanding guides is perhaps the next best thing.

I was overwhelmed by the richness of life as a boy and stimulated by the relatively few children's books available on natural history in the 1940s and 1950s, but if I had had the good fortune to encounter this book at an early age, I would have been delighted—but no more so than I am now. Turning the pages, I was reminded of Jean-Henri Fabre's nineteenth- and early-twentieth-century essays on the life histories of insects in southern France, which fascinated me; of Charles Darwin, John Muir, and Aldo Leopold; and of many other fine writers and philosophers. It is my opinion that anyone with the slightest interest in biology will devour the pages of this book, as I have done, and will come away satisfied.

The world in which we live today has changed drastically from the one into which we were born. I live in a world in which there are 3 people for every 1 who was alive in 1936; one in which 250,000 babies are added to a global population of 7.4 billion every day; one in which we are consuming some 150 percent of the sustainable productivity of our planet (see http://www.footprintnetwork.org) on an ongoing basis; one where 800 million people are malnourished, with Oxfam estimating that 62 people control half of the world's wealth while half of the world's people are malnourished; one where some 200 million people have been killed in wars over the two centuries since the world was fully divided into distinct nations; and one where nuclear arms, with the United States having seven thousand and Russia another seventy-seven hundred nuclear-tipped missiles, are spreading to additional countries, rich and poor.

Global climate change, substantially driven by human activities, is altering the very conditions in which life exists. The pine processionary caterpillars whose long head-to-tail trails formed the subject of a charming work by Jean-Henri Fabre, published in 1898, have now followed gradually warming climates up into England, something that would have been thought impossible a century ago. The climates of the fields where corn and soybeans are grown so successfully in central Illinois are experiencing drastic change, so that the whole future of agriculture in this highly productive region will need to be rethought and redeployed over the next few decades if we are to avoid still further starvation.

Not only are we driving species of all kinds of organisms to extinction at a rate that has not prevailed for some sixty-six million years, but we do not even know most of them, much less the kinds of unbelievably beautiful and intricate relationships of which we are presented such a fine sample in this book. Other than bacteria and archaea, for which we cannot even provide a meaningful assessment of their diversity, there are probably at least twelve million kinds of organisms existing on earth now, and a substantial proportion may be lost by the end of this century—possibly even half.

We depend on other life for our lives, but this book provides clear insights into another reason that we depend on it—for satisfying our insatiable curiosity, our hunger for beauty, and our desire to create beauty. If we allow our Stone Age greed for accumulating as much as we can and consuming it to destroy us in an age of belligerent, self-seeking nations, we will have nothing to blame but ourselves. The civilization that we have built, largely during the most recent 10,000 years of our 3.4 million year history on Earth, is precious. It produces music, and art, and philosophy, and fine writing like that exhibited so well in the book you are reading now. Perhaps the insights and reflections presented

A luna moth, one of the 160,000 species of moths found worldwide (*right*).

here will help to convince us to do better; to love one another, and the Earth that supports us, more than we do; and to find a way to produce a sustainable future for ourselves and our species. In any case, it certainly will have filled us with pleasure and inspiration along the way.

—Peter H. Raven, President Emeritus, Missouri
Botanical Garden, St. Louis

ACKNOWLEDGMENTS

Any time a work encompasses more than a forty-year time span, remembering to credit all the people who helped along the way becomes extremely problematic. Our memories simply are not capable of pinpointing each and every individual, so if anyone has been left out, it is our oversight.

This manuscript would not have been possible without our adventures, many of which were made possible by our friends, colleagues, and a very special tour company—Tropical Birding—that has opened up the entire world for our exploration. For our Illinois adventures, we thank Liz Jones of the Cypress Creek National Wildlife Refuge and her Renaissance husband, Dave Shaffer, for providing easy access to many of the best places to explore in southern Illinois. Jim Waycuilis and Molie Oliver opened up the Cache River State Natural Area to us, and Jim accompanied us on many forays into this unique ecosystem. Jim Wiker accompanied us on trips involving butterflies and moths, and we are grateful for his encyclopedic knowledge of Illinois lepidoptera. We owe special thanks to Dr. Robert Reber, who served as editor of *Illinois Steward* magazine for many years and organized our visit to "the Shack." From the other side of the globe, our friend and colleague Dr. Raghu Sathymurthy aided us in our Australian sojourn. We thank the University of Illinois's Osher Lifelong Learning Institute (OLLI) for choosing us to lead a trip to Alaska for their members. Otherwise, the Flying Squirrel Café would be only a dream.

Many of our friends and colleagues from the University of Illinois's Prairie Research Institute (PRI) provided much-needed information and expertise for the analysis of many of our observations. Connie Carroll Cunningham is a botanist extraordinaire; James Angel, the Illinois state climatologist, helped demystify certain weather-related phenomena; Dr. Michael Dreslik lent his herpetological expertise on several occasions; Dr. Rod Norby and Michael Jared Thomas provided fossil trilobites for us to photograph; Charlie Warwick edited portions of the manuscript; Dr. Greg Sass put us in "intimate" contact with Asian carp; Dr. Ed Zaborsky alerted us to crazy worm circles; Chris Grinter identified bizarre caterpillars for us; Paul Marcum pinpointed the locations of rare orchids; and Patty Dickerson lent both her sketching and her photographic talents to complete several essays and scanned numerous images for the manuscript.

Over the years, our interactions with Larry Gross and James (Mac) McNamara, both talented staff members at the Illinois Natural History Survey (INHS), have led us to attempt many enterprises that would have been impossible without their expertise. Our interaction with southern Illinois botanist and friend Karen Frailey always provides an added dimension of "enthusiasm" to our experiences. We owe a special debt of gratitude to Dr. Phil Mankin, wildlife biologist and pilot. Phil never failed to show us a unique perspective on the world, all the while keeping us safe at two thousand feet above the landscape! We express our gratitude to Bob Bunselmeyer for sharing his very odd soybean-harvest experience with us. For mammal questions, we relied on Natural History Survey scientists Dr. Joseph Merritt and Dr. Joyce Hofmann, while Dr. Ed DeWalt provided insight into the lives of winter stoneflies. Retired scientist Dr. David Thomas and his wife, Carol, granted access to their home swimming pool for a "photo session."

We also thank Thomas (Tad) Boehmer and the staff of the University of Illinois Library, Rare Book Room, for finding and imaging portions of rare manuscripts. From the University of Arkansas, Dr. Robert Wiedenmann invited us to accompany him to Mexico to experience monarch overwintering. Rob also introduced us to the magic of "birding travel." Our special gratitude goes to graphic artist and biologist Danielle Ruffato, who applied her extraordinary talents in designing this book. Sue has a special thank-you for Dr. Donald Kuhlman of the University of Illinois for his encouragement that led to her career in entomology and for physical therapist Kim Hardin, who introduced her to a training regimen directly applicable to pelagic birding. We also acknowledge Mark Bee, protozoologist, who helped put our observations on nature into perspective, and INHS entomologist Charles Helm, who not only is a good friend but also mentored us during our early years in science. Zack Lemann at the Audubon Butterfly Garden Insectarium answered relevant questions regarding katydid coloration.

Many of our overseas adventures were made possible by the guides and staff of Tropical Birding. We offer our most special thanks to guides Iain Campbell (just being with Iain is an adventure!), Keith Barnes, Pablo Cervantes Daza, José Illánes, and special friend and guide Ken Behrens. Their worldview has become our worldview, and the experiences we have had with them are priceless. On a more intimate level, Lark, our guinea pig, would always *wheek* with delight after each finished essay, as she knew a treat was forthcoming!

We are deeply grateful to the following individuals at the University of Illinois and PRI who served as our scientific editors on this manuscript: entomologist Dr. Joseph Spencer spent countless hours copyediting—which enormously improved the manuscript—and Joe also contributed his photographic and cartoon-creation expertise. Paleontologist Dr. Sam Heads applied his wide-ranging biological and geological knowledge to the work, and entomologist, artist, writer, and naturalist Dr. James Nardi served as the final reviewer to make sure we had dotted every scientific

i and crossed every *t* in the essays. We offer special thanks to Dr. Bonnie Styles, director emeritus of the Illinois State Museum, for her technical review of the manuscript and to Dr. Peter H. Raven, director emeritus of the Missouri Botanical Garden, for contributing the foreword for this work. Luann Wiedenmann was a delightful traveling companion and always seemed to have that perfect phrase that put our adventures in perspective. We would be remiss not to acknowledge two gentlemen scientists—Dr. Robert Metcalf and Dr. James Sternburg—who now explore worlds beyond ours. Their mentorship at the University of Illinois led us to a lifetime of biological exploration and discovery.

To be able to document a collection, whether it is objects or experiences, requires a reference library, and we were fortunate to have access to a gem—the Illinois Natural History Survey Library. Here was housed a treasure trove of field guides and natural history references. During our tenure there, the librarians were always ready to find an obscure reference for us or order a new field guide they thought would be of interest. Sadly, like several of the observations, places, and organisms featured in this book, our library is no more. This wonderful collection of books and knowledge that led to countless opportunities will be missed.

Finally, we thank our parents. While they may not have always under-stood our individual and collective passions, they never stood in the way and were always encouraging us to pursue "what made us happy." As it turned out, exploration and discovery send us to our "happy place," and we will continue to observe and explore nature, as there are always new adventures just across the road.

Iain Campbell, our guide from Tropical Birding, takes a break from his very demanding clients (*below*).

INTRODUCTION

NATURALISTS AFIELD

A LIFETIME OF CURIOSITY

Almost every day... I usually made for the fields. ...
My little basket went with me... and came back full of
bird's eggs, nests, lichens, flowers, and pebbles.

—J. J. Audubon, *Audubon and His Journals*,
vol. 1 (1897)

HISTORIC CABINETS
OF CURIOSITIES

Humans—the "pack rats of the primates"—have always been collectors. For much of human existence, our "collections" consisted of stores of foodstuffs, materials for weapons and tools, and other objects that made life, if not easy, at least possible. However, beginning in the seventeenth century, the world changed as humans developed the technology to explore our exceedingly large and diverse planet, and with it came an eager appetite for the earth's artifacts. Explorers sailed away and brought back wonderful things from across the globe. Although unicorns, mermaids, and other mythical creatures were not part of the mix, the objects these explorers did bring back were equally as enthralling.

The main collectors during the seventeenth century were the wealthy archdukes, kings, and emperors of Europe. They could afford to indulge in the luxury of nonsubsistence collecting. Patrick Mauries writes in the introduction to his book *Cabinets of Curiosities*, "Their motives were fundamentally the same, 'to condense a whole library into a single book.'" He goes on to state that the seventeenth and eighteenth centuries were "the last period of history when man could aspire to know everything." In other words, this thirst for objects was viewed as a search for universal

knowledge. For these individuals, collecting often became an obsession, and the space needed to house their collections ranged from simple cabinets to overcrowded rooms and entire mansions. Many of the collections were narrowed in some fashion; after all, it was not possible to collect and accumulate everything. Individual cabinets often had themes or foci around which the artifacts were amassed. Individuals might specialize in minerals, fossils, stuffed birds, or anything else that struck their fancy.

Collections of objects from the seventeenth and eighteenth centuries typically had thick leather-bound books that detailed where the artifacts originated and any other information deemed relevant. Today, we categorize the information that accompanies artifacts as provenance—a documented history of the item. *Provenance* is now a common term, thanks to the Public Broadcasting Service's *Antiques Road Show*. Collections of things—even simple things—acquire power from the mere act of accumulation. Additional information only increases their power, and objects with a provenance immediately have greater value and increased significance. For example, there can be no more common, mundane object on the face of the midwestern landscape than a dried corn shuck. However, add provenance and perhaps the passage of a certain amount of time, and even a dried corn shuck can achieve interest, significance, and, ultimately, value. We (the authors) witnessed the final harvest of a cornfield that was to become the Nature Conservancy's Emiquon Preserve in central Illinois. The Nature Conservancy had purchased the seven-thousand-plus-acre property (the former bed of Thompson Lake adjacent to the Illinois River), and the process of restoration began after that last harvest. We collected one of the last shucks of corn from the property, just

before it entered a dusty combine. Today, as ducks and geese darken the skies over Emiquon, that corn shuck represents a change in the fabric of history.

MODERN CABINETS
OF CURIOSITIES

While few of the unique, old-time cabinets of curiosities survive today, an abundance of engravings and other images of these collections do live on to pique our interest. But do cabinets of curiosities still exist in some form? The answer is a resounding "Yes!" Over time the objects that accumulated in these cabinets ultimately provided the foundations for what we now call museums. For example, many of the artifacts the participating countries left behind from the 1893 Columbian Exposition in Chicago formed the base collections of the Field Museum.

On a more personal level, we have both spent our careers employed by the Illinois Natural History Survey, now part of the University of Illinois's Prairie Research Institute. The INHS did not always have such a lengthy or lofty title. Its story began in pre–Civil War Illinois (1858) as the Illinois State Natural History Society. In the *Transactions of the Illinois State Agricultural Society, 1859-60,* a short article titled "The Museum of the Illinois State Natural History Society" notes, "The State Normal University [Bloomington, Illinois] has been made the depository for all collections in the various departments of Natural History which may be made under the auspices of the Society, and also for such collections as may be donated." Based on historical images, the collections of the Illinois State

Natural History Society appear remarkably similar in content and style to a European cabinet of curiosities.

Later in the century, in 1877, the museum became the State Laboratory of Natural History, and Stephen A. Forbes was named its director. In 1917 the State Laboratory of Natural History and the Office of the State Entomologist were merged to become the Illinois Natural History Survey, and Forbes was selected its first chief. The organization was moved from Bloomington to Champaign, on the campus of the University of Illinois, where it resides today. A visit to the present-day survey reveals little of its cabinet-of-curiosity beginnings, unless you are

Seventeenth-century Ferrante Imperato cabinet of curiosities from Italy (*above left*; image courtesy of University of Illinois Library, Rare Book Room). Museum of the Illinois Natural History Society, 1859 (*above right*; image courtesy of Illinois Natural History Survey).

privileged to tour its extensive biological collections. Unlike the Smithsonian, which has been dubbed the "nation's attic," the survey's collections represent the collective "biological memory" of the state of Illinois. When a visitor enters the various collection rooms, no great trumpeting elephants or imposing dinosaur skeletons startle him or her with their presence. The survey is not a public museum but a research organization with large collections. Even so, visitors are often struck by the very "un-attic-like" sense of impeccable organization—a place for everything, and everything in its place.

THE ORIGIN OF OUR
CABINETS OF CURIOSITIES

Not surprisingly, we (the authors) have always been collectors. Because of our respective and collective passions, it should come as no shock that we now reside in our own "cabinet of curiosities." Our friends often note that our museum-like home seems relatively unlivable, has too much stuff, and must be a nightmare to dust. Of the three, only the last phrase is true! We cherish our collections, use them for endless sources of inspiration, and are continually on the lookout for those perfect additions.

We both have come by our devotion to collecting from a combination of nature and nurture—not surprisingly, though, by separate, but remarkably parallel, roads.

MICHAEL JEFFORDS—Growing up in a very small town in the tip of southern Illinois, I had an abundance of places nearby to entertain me—the Ohio River and its backwater lakes and sloughs, the Shawnee National Forest, the "largest railroad-tie yard in the world" immediately adjacent to my house, and a host of other magical places. From the age of five or six, I was basically allowed to roam free. When my mother would say, "Go out and play, but be back in time for lunch [or supper]," she had no idea what type of "play" and license to roam she had unwittingly endorsed. Her only remonstrance was the inevitable "Don't go near the river; it's too dangerous!" Of course, that just made the river all the more enticing. Often when I returned home covered in mud and reeking of decaying fish from a day spent exploring flotsam, jetsam, and other treasures along this massive stream, she would inquire, "Have you been to that river?" My answer was always, "Nope, just down to the spillway." "Okay, then get cleaned up for supper." The fact that she had never

A portion of the Illinois Natural History Survey's insect collection (top left; photo by David Riecks, courtesy of Illinois Natural History Survey). A small portion of the authors' cabinet of curiosities (bottom left).

been to the spillway and had no idea that it was a short stream immediately adjacent to the river served only to rationalize my little white lie.

As I grew older, bolder, and stronger, my parents soon began to realize that my roaming was more than just "going out to play." One day I discovered a nearly complete horse skeleton and conjured the idea of bringing it home to reassemble in my budding natural science museum (my collection was slowly overtaking our rickety back porch). As the skeleton was located nearly a mile away in a forested ravine, it was quite a task for an eight-year-old to haul it home; however, I managed most of it, piece by piece. My father was somewhat taken aback by this but allowed the bones to accumulate in our backyard until I had a reasonable facsimile of a full-grown horse. I never did manage to completely reassemble it; I lacked the technology, as Elmer's school glue didn't quite do the job.

Even though my father claimed to not understand my fetish for bringing home such treasures, he was likely the source of this genetic tendency. Although from a generation that worked themselves seemingly to death to support their families in poor, rural southern Illinois, he loved to "go to the woods" and would regularly disappear for many hours on a Saturday. Late in the day, he would return with a sturdy cloth sack literally bulging (often upwards of seventy-five pounds) with pecans he had spent the day gleaning from the same riverside forests that I explored. Oddly, we never went together, and I actually never encountered him in the woods. It obviously was his time away from the toils and travails of working and marriage (by the time I was old enough to notice, my parents no longer got along very well, and I grew up in a basically divided home). Now, I realize that his generation could not merely enjoy nature for nature's sake; rather, they had to come back with some "product" to justify spending that much time in nature. Since he did not hunt or fish, the pecans were his justification. My uncle who lived down the road from us was much the same way, although the "reason" for his nature activities involved ginseng collecting and trapping to supplement a meager income. As with my father, I never met my uncle Frank in the wilds of southern Illinois, but he and I

often spent hours discussing his adventures.

My passion for nature was soon channeled from all things collectible into what was to eventually become my profession—entomology. At age nine, I encountered a giant female Cecropia moth on my way home from school one spring day. That one event triggered an almost insatiable appetite for insect collecting, which continued over the next twenty-five years. I built my first insect boxes from blue Styrofoam and literally plastered the walls of our house with case after case of Illinois insects. For a time, these insect boxes created quite the sensation, and folks often visited with my

mother for a smoke and to enjoy my collection. Early in my collecting enterprise, I inadvertently discovered a catalog devoted to the sale of "worldwide" insects from a dealer in Florida. Most of the prices were far above my allowance, but a short sentence caught my eye—"I will trade foreign material for unusual insects from the U.S." This immediately piqued my interest, and I soon found the perfect way to supplement Illinois insects with spectacular examples from across the globe. Page 2 of the catalog listed the "rare and unusual elephant stag beetle from the eastern U.S., for $60/pair." As it happened, I regularly collected this

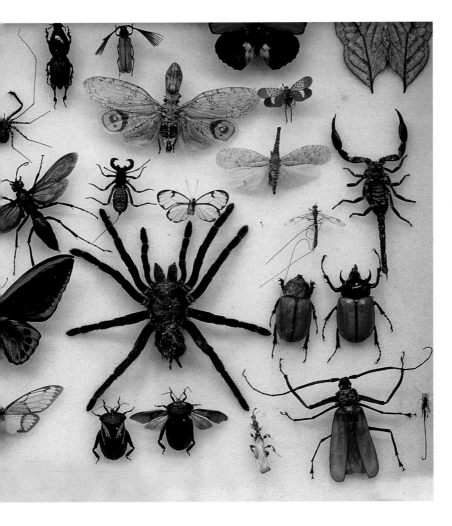

Nothing much has changed over the years, as Michael Jeffords proudly displays his Christmas gift (a horse skull) from his in-laws (*far left*). Sample of traded specimens from around the world in Michael Jefford's collection (*left*).

"rare" insect at the lights of the local coin laundry. It was a common insect in southern Illinois, and over the next several years I was able to trade many pairs of stag beetles for examples of the world's most spectacular insects.

While I maintained my collecting passion, over time I became less of a collector and more of a student (literally) of nature. Something was missing, though, and during my undergraduate college years (the late 1960s and early '70s), I was somewhat at a loss as to where my real passions lay. I always carried a nagging thought buried deep in my brain that there must be

another way to "collect" nature without having to cause any creature's demise.

After an ill-fated stint in medical school, I dropped out and entered graduate school at the University of Illinois in entomology. Finally, I had found a home. It was in graduate school that I discovered the mechanism I had been seeking all along to portray (and collect) the natural world. This endeavor did not involve artifacts or insects on pins; it simply required film and a camera. My neighbor Mary O'Keefe helped me get started, and it just so happened that my thesis adviser was an accomplished insect photographer.

What more could I want? Over the years, my skills as both an entomologist and a photographer developed, but one thing was missing: a partner in this lifelong quest was needed, and one soon materialized.

SUSAN POST—I write this while sitting at an oak library table that once belonged to my great-grand-mother. I purchased it at a cousin's auction while Michael sat under a tree reading a book—oblivious to the people and auction patter. His only comment was "Did you get it?" The library table was a wedding present to ourselves, as we were to be married a month later; antiques were not his thing!

Collecting is part of my genetic makeup. My great-grandmother Olive Hallam Webb was a collector; she donated eighty-two items from her collection to help furnish the Herndon Cabin at New Salem State Park, Illinois (famous for where Abraham Lincoln developed as a man and as a statesman), and a set of Westward Ho dishes to the Smithsonian Institution. Although I never met my great-grandmother, I still felt that I knew her. When visiting my grandma Babe's house—my great-grandmother Olive's daughter—Grandma would often recount stories about the unique items in her home and the associated history (provenance) for each one.

Other early memories are of accompanying my father to local farm auctions. My mother, a registered nurse, worked many weekends, so my father would take my siblings and me to these auctions. Yet the first sale I really remember was with my mother. She purchased a complete pitcher and bowl set in a blue and white pattern. It was the late 1960s, and it was her first auction buy. I believe she paid twenty-five dollars for the set, and everyone around her thought she was crazy!

A seed picture by Grandma Babe that resulted from numerous trips to the woods (*left*).

I grew up in the country in central Illinois. We did not farm, but our house was surrounded by agriculture. Across the road from our house was a "woods." The woods—which consisted of a few scattered oak and hickory trees and a creek—was actually a remnant of an old prairie grove. It was used as a cow pasture, but to my siblings and me it was magical. When my grandma Babe would visit (which was at least every other week), we would beg her to take us to the woods, as we were not old enough to cross the road on our own. She would lead us into the trees, where we would look for wildflowers, nuts, and leaves. Although my grandmother had never finished high school, she was a wealth of knowledge and would tell us stories and sing songs about the plants and animals we found. We always managed to fill her apron pockets with nuts and leaves, which Grandma would take home and use to create her signature "seed pictures."

Growing up in a small agricultural community during the 1960s and '70s, everyone joined 4-H, sponsored by the county Cooperative Extension Service. I was no different, but this was before the passage of Title IX legislation; there were activities for girls (cooking and sewing projects) and activities for boys (animal and gardening projects). Of course, I wanted to participate in the boys' activities, but first I needed a project. When my mother suggested entomology, saying that it sounded interesting and I might learn something, my first response was the expected, girly, "EEW! BUGS!" After all, I was only eight years old. But my desire to join the boys' club outweighed the "yuck" factor, and soon our house was littered with jars, as one of the requirements of the project was an insect collection. The first year required twenty-five insects, with twenty-five more added each subsequent year of participation. I was in 4-H for ten years and had entomology as a project for nine of them.

I roamed my little corner of nature—our two and a half acres and the woods across the road—looking for insects. My siblings were my partners, carrying nets, jars, and bags. While the woods and creek across the road was technically still off-limits, we would regularly sneak over there to collect. Of course, when asked if we had been to the creek, our answer was "No, of course not!" Yet I am sure we never fooled anyone, as we always managed to fall in and carried home a distinct odor. My youngest sister recently reminded me of our entomological exploits, as she had found a picture she had drawn in the first grade of us examining cow pies for insects. This could have been another source of our unique smell!

Over the years, my insect collection and my knowledge of entomology grew. Even during a summer abroad in northern Germany as an exchange student, I collected insects. The family I lived with bought me a bicycle, and I roamed the countryside looking for new species to add to my collection; it was something familiar in a country where I did not speak the language. Imagine the chaos at U.S. customs when they asked if I had anything to declare. "Just this box of insects," I said. I knew nothing about permits then. Someone shined a flashlight at my neatly pinned and spread specimens, confirmed they were dead, and waved me on.

I began my freshman year at the University of Illinois as an animal science major, with the goal of becoming a veterinarian. However, this aspiration was short-lived and faded once I had a "taste" of animal physiology. What should I do? It was at that point that I decided to make an appointment with Dr. Donald Kuhlman, an agricultural entomologist at the Illinois Natural History Survey. Dr. Kuhlman had judged my 4-H insect collection at the state fair and had written a letter that I brought along, congratulating me on my A rating. In the letter, he had stated that if I ever had any questions to look him up. During my meeting with him, he gave me advice on what some of the career options were for entomologists and what classes to take.

While the University of Illinois did offer an undergraduate entomology degree in the curriculum, what the course handbook did not tell me was that it was discouraged, and entomology "should be" reserved for only advanced degrees (master's or Ph.D.). Fortunately, this small detail did not deter me, and beginning my sophomore year I was majoring in entomology. One of my first classes, recommended by Dr. Kuhlman, was Agricultural Entomology. I fit right in, and while my teaching assistant (Michael Jeffords) somewhat scared me, I loved the class.

During my final semester at Illinois, some of my friends and I went to a photographic art show at the Illini Union Art Gallery called *A World of Insects*. The artist was former teaching assistant Michael Jeffords. On the walls were framed insect specimens and close-up photos of insects. I remarked to my friends that I sure would like to be able to take photos like that. Who knew that six months later Michael and I would be roaming the countryside together? My grandma Babe had given me money to buy a 35mm camera for graduation, and I had recently purchased a car. Michael and I both ended up working for the same person at the Illinois Natural History Survey, on soybeans, no less. Michael had camera equipment I could borrow, and he was willing to teach me photography. I had a car, whereas he used the bus, so we struck up a friendship based on mutual need at the time. During 1981 two events occurred. As Michael and I began to explore Illinois together, one thing became obvious: I had to develop my own "photographic eye." Michael's specialty was macro photography (close-ups) of insects. I could not just emulate him. While I dabbled with insects, I soon gravitated toward plants. Having had a field-botany class in college, I was familiar with some of the families, but not nearly knowledgeable enough. To identify and learn the plants, I began to carry a small three-by-five-inch notebook with me. With field guide in hand, I identified the plants and recorded them in my notebook. Soon my notebook entries went beyond the plant names and included growth stages—were the plants just leaves, in bloom, or setting seed? As we continued to roam Illinois, these notebooks were soon filled, not with just botanical information but also with bird, insect, and the occasional Illinois mammal observation.

Unfortunately, my grandma Babe had a stroke during 1981. She would be confined to a nursing home, and while she could visit our home, long excursions outdoors with her were no longer an option. Using my notebooks, I began to write weekly letters to my grandmother at the nursing home. I would describe what I saw, using the words of Rachel Carson as my guide. "One way to open your eyes is to ask yourself, 'What if I had never seen this before? What if I knew I would never see it again?'" My grandmother not only read my letters but would then share them with everyone at the home.

I still carry a three-by-five-inch notebook wherever I go and consider my collection of more than one hundred of them as one of my most valuable possessions. Here reside the notes and observations of everywhere we have traveled, locally, regionally, or globally. Some pages are warped and faded from rainstorms or hail events. On the back of one notebook is taped a penguin feather, tracked in after a day spent roaming an island in the Falklands—just the penguins and us. Others are sweat stained from birding adventures in tropical rain forests. When someone asks how I am inspired to write, I paraphrase poet Mary Oliver. My three-by-five-inch notebooks "are the pages upon which I begin." Today, Michael and I still explore the world together, and after thirty-plus years of marriage, the adventure is still just across the road.

OUR LIFELONG CABINET OF CURIOSITIES

We are biologists—entomologists to be exact—who have made it a part of our lifelong objective to photographically document the natural world. Over the past forty-plus years, we have amassed an impressive collection of almost a half-million images taken in Illinois, the United States, and, during the past several years, the world. We have photographed in most areas of North America

Caspian terns roost on Michaelmas Cay, Great Barrier Reef, Australia (*above*).

and explored some of the most biologically diverse areas on Earth—Manu National Park in Peru, the cloud forests of Ecuador, southern Africa and the horn of the Okavango Delta, the Falkland Islands, the Pantanal wetlands of Brazil, and northwestern Australia. We are not photographic trophy hunters, seeking to make each and every image that perfect portrait, but biologists who seek to capture the little windows of time that show nature as it is. Don't get us wrong: we do strive to make the best images possible, but often modified by the set of circumstances presented to us. For example, one of our favorite sets of photographs comes from the South China Sea, off the coast of Australia, at the edge of the Great Barrier Reef. We and three other folks were circling Michaelmas Cay in a rubber dinghy, bouncing in the heavy surf, all the while trying to deploy handheld long lenses at the plethora of birds arrayed along the shoreline before us. All the

images are certainly not perfect, but they definitely serve the purpose of showcasing the remarkable creatures as they gathered on that remote dot of land.

Our photographs are stored as slides and housed digitally as many terabytes on computer hard drives. This collection has allowed us to create exhibits, educational publications, books, and innumerable other materials associated with our work as scientists. However, our massive photo collection is not the reason for this book. Rather, it has a narrower focus that harks back to those seventeenth-century cabinets of curiosities.

Over the decades that we have been "collecting" images of the natural world, we have happened upon a finite number of observations and experiences that we view as unusual in the extreme. These observations

and experiences—we have whittled down the list to 102—are the focus of this book. Some are one-time events; others are stories put together over many years of observation. Each two-page spread features a short essay that explains the circumstance and, hopefully, interprets the science behind each event. Some of the occurrences can be explained with existing scientific evidence, while others are less obvious and require a fair bit of speculation to accurately interpret what we are observing. While most are illustrated as simply as possible with our own photographs, there were a few occasions when we actually missed the once-in-a-lifetime shot; consequently, those observations feature illustrations

King penguins heading to their colony on the Falkland Islands (*below*).

created with the help of computer software and even the occasional cartoon drawn by Dr. Joseph Spencer, a friend and colleague from the INHS. We have also included a glossary of terms that may be unfamiliar to the general reader.

To us, this is a lifetime of observations distilled into a single work. As scientists, our perception of the world likely differs from that of most casual observers. The experiences we have chosen to showcase range from encounters with unusual natural history phenomena in our own backyard to observations from the remote corners of the earth. We hope you enjoy our intellectual and experiential cabinet of curiosities.

AMASSED NATURE

ver since reading about the sky-filling aggre-gations of passenger pigeons, we have been searching for nature in numbers. Maybe it is because, as collectors, one or two of anything seems so lonely. Over the years, we have seen our share of the great wildlife spectacles of the world—thousands of pen-guins sprawling across the rolling green sward of islands in the southern Atlantic, great herds of Cape buffalo and wilde-beest marching across the African plains, even masses of flamingos in the shallow waters off the Yucatán peninsula of Mexico. We have also missed many of the signature wildlife events showcased by the media, but have encountered additional obser-vations that lie outside the realm of "the rich and famous," largely "off the radar" of seekers of amassed wildlife. Many of the observations we have chosen to depict in this section are less familiar, but no less relevant, than those anyone would expect to see from a chapter titled "Amassed Nature." Many are just on a smaller scale. These encounters have resulted from far-flung adventures and sometimes from everyday "mundane" events—traveling to central Mexico to stand amid a hundred million monarchs, watching a New Mexico sunrise over an ever-whitening multitude of snow geese and witnessing their exodus against a rainbow sky, hearing the seething masses of a peri-odical cicada emergence in our local woods, or just looking closely at a writhing clump of mites clustered at the back door of our office building.

A Cape buffalo herd heads to the Sand River, Kruger National Park, South Africa (*previous page*). Flamingos gather in massive numbers at Ria Lagartos, Yucatan, Mexico (*left*).

These experiences, and more, contribute to the richness of life. They have inspired, awed, and sometimes cowed us with our attempts to portray these natural wonders. While some opportunities to encounter them recur on an annual basis, some events happen rather sporadically, and others may even be once-in-a-lifetime occurrences. When this happens, the pressure is on! If we fail to capture these marvels of nature, they will still become part of our rich trea-sury of personal memories, but memories will be all that we have to share. Without some sort of physical expression of what we have seen and experienced, we will be unable to communicate the phenomenon and risk the danger of having fewer people know, under-stand, or care about nature and its intricacies. Here we present samples from a lifetime of observation, some subtle, some not so subtle, others minuscule, but all important manifestations of nature and the natural world.

TURTLE SHINGLING

MICHAEL JEFFORDS

Nature often provides great models for human activities. A Web search for "turtle shingling" yields a bevy of roofing tips on various vent and ridge-cap techniques. The phenomenon that gave rise to these roofing practices may not be familiar to most, so it is worth a closer look. The red-eared slider, a common turtle (order Testudines) of eastern North America, is most familiar as the "cute little babies" sold by the pet trade. Though native to the southern United States, it has become established in many other areas by releases of these "pets" into the wild. The slider is now considered an invasive species that can outcompete other native turtles. When a turtle is observed in the wild, often sharing a log with several of its fellow Testudines, it is usually a red-eared slider. Most of us have observed turtles lined up on a log in precise fashion, basking in the sun, usually while we whiz along on a busy highway. We take little notice, but if we do stop to observe, the turtles have seen us and *plop, plop, plop...* the only behavior we see is the escape response.

Red-eared slider turtles sunning on logs.

Why do turtles bask? An overriding reason is thermoregulation. Turtles are poikilotherms—they cannot regulate their internal body temperature and are completely dependent on the temperature of their surroundings. For this reason, they must constantly sunbathe to warm themselves and maintain their body temperatures. If the body temperature falls below 21°C, their digestive processes cease to function. The intriguing term *leech-load* refers to another reason for basking. Leeches are common parasites of turtles, and the leech-load can range from zero to more than one hundred leeches per turtle (depending on the size of the turtle)! They often attach under the shell to protect themselves while turtles are basking. Too much sun and the leeches dry out—certainly a benefit for the turtles.

While basking behavior of turtles has been widely studied, relatively little is known about the individual interactions when these turtles come together. Remember, observations are usually limited to a fleeting glimpse, followed by a mass exodus from the favored log. However, various researchers (through patient observation) have noted that while basking on logs, turtles exhibit various aggressive behaviors—biting, pushing, mouth gesturing—that are thought to maintain optimum spacing between turtles and allow maximum exposure to the sun. When turtles become numerous enough that they overlap (or shingle), a "lateral displacement" maneuver may occur. The turtle on the bottom of the shingle stack rocks side to side, apparently to dislodge an offending neighbor. This may cause a wavelike motion that ultimately ends up affecting the entire cadre of turtles. As you might expect, the larger the turtle, the more impact its lateral displacement will have on the turtle stack. While seeing turtles "do the wave" is seldom observed, just observing large numbers of stacked turtles is reward enough. My personal record for turtles shingled on a log is fifty-four individuals! It seems the only limiting factor is the length of the chosen platform.

HACKBERRY HAPPENINGS

SUSAN POST

Mass spectacles of butterflies usually refer to the monarch aggregation in the mountains of Mexico, but this is not the only butterfly extravaganza the world has to offer. The hackberry butterfly, a small, brown North American species, often noticed only when it lands on a sweaty arm, can also exhibit massive gatherings. Unlike the showy monarch, hackberry aggregations are unpredictable, and adults are on the wing for just a few weeks. Nevertheless, their outbreaks are still almost unbelievable! The sheer biomass of hackberry butterflies decorating the sky, ground, and any other object is a source of wonder during early summer. Beginning in 2009, we traveled throughout Illinois with two goals: to capture photos for a new field guide on Illinois butterflies and to increase our knowledge on the status of the state's butterflies. A most curious event that year was the appearance of a multitude of hackberry butterflies. My field notes from Sand Ridge State Forest on June 12, 2009, read, "Walking through a cloud of butterflies as they 'puddle' on the road at sap, and on the bodies of their fallen brethren, I can hear their wings—a crinkly paper sound. Thousands adorn the highway after a synchronous emergence." The sheer numbers reminded us of our visit to the monarch overwintering grounds in Mexico. The hackberry butterfly is found wherever its caterpillar food plant, hackberry, grows. The adult rarely visits flowers, preferring sap, rotting fruit, carrion, dung, or wet spots along streams or roads. While the species frequently rests high in trees, it is also salt loving, and adults will often land on the arm or clothing of a perspiring observer, allowing a close look. The female lays her eggs, singly or in small to medium clusters, on young hackberry foliage. The partially grown caterpillars overwinter in rolled leaves but disperse to feed in spring.

Hackberry butterfly numbers vary greatly from year to year. Was this emergence at Sand Ridge unique, or had there been other occurrences? Yale Sedman and David Hess documented two different swarms in their book, *Butterflies of West Central Illinois.* "Populations of the hackberry butterfly may become enormous so as to reach the levels commonly seen in economically important species. On June 13, 1978, near Rockport, in Pike County... [hackberry butterflies] were seen in immense swarms covering miles of roadway." An island in Lake Erie in the 1950s had so many hackberry butterflies that "the odor of crushed leaves could be smelled by people on boats passing offshore." Within four years, the caterpillars had killed the local hackberry trees. For the past few years, Douglas County, Kansas, has experienced an outbreak of hackberry butterflies. They cover buildings and form dark clouds in the sky, and the local Chamber of Commerce invites visitors to this amazing butterfly phenomenon. The cause of these population explosions is unknown. So next time you need a butterfly-spectacle "fix," perhaps a visit to the "mountains" of Illinois or Kansas is just the ticket.

Hackberry butterflies cluster on a rural roadside in western Illinois. The grouping forms when a few individuals stop to feed, become victims of passing cars, only to attract more of their brethren.

JAPANESE INVASION

MICHAEL JEFFORDS

As an entomologist, environmental issues are never far from my thoughts. Such things as exotic invasive species, misuse or overuse of pesticides, and the worldwide decline of pollinators trouble me. These monumental themes threaten the integrity of the biosphere and directly impact our daily lives. Seldom are all of these issues embodied by a single

The introduced Japanese beetle can explode during a given year and reach enormous population levels.

organism. The Japanese beetle (*Popillia japonica*) is such a creature. Introduced into the United States prior to 1912, it was first discovered in a nursery in New Jersey in 1916 and has become a serious pest of more than two hundred species of plants since then. In Japan the beetle is controlled by natural predators, parasites, and diseases and is not a problem.

My own experience with the Japanese beetle began in graduate school at the University of Illinois during the class Insect Pest Management. One assignment was to read *Silent Spring* by Rachel Carson. Here we learned of an all-out war that was waged against these beetles in the Midwest, best epitomized by the events in the small Illinois town of Sheldon. Beginning in 1954 and continuing until 1961, Dieldrin (a broad-spectrum organochlorine insecticide) was sprayed over the area at the rate of three pounds per acre, resulting in enormous losses of birds and other

wildlife. The Illinois Natural History Survey, my place of employment for thirty-five-plus years, documented the devastation. Ironically, I started my career there as a soybean entomologist, and during the 1980s I was called to sample soybean fields near Sheldon. Upon arrival, the problem soon became apparent, as one hundred sweeps with my insect net resulted in a bag full of Japanese beetles that weighed more than three pounds! Needless to say, the earlier eradication attempts had failed, and outbreaks of Japanese beetles still plague the Midwest. Today, as a home owner and a gardener, I dread each summer's emergence of the beetles, as they can devastate even the most well-tended gardens.

But what does all this have to do with the decline of pollinators? While we often associate the invasion of exotic invasive organisms with a decline in overall biodiversity in the invaded landscape, it is often difficult to pinpoint the physical interactions that actually drive specific declines in species, such as pollinators. I have been "privileged" to witness the effects of Japanese beetles in native Illinois prairie habitats during the past few years, and the story is a sad one. During several years of work on a field guide to Illinois butterflies, my favorite butterfly sites—prairies full of blooming milkweed and other species—were often bereft of butterflies. Japanese beetles were so abundant that they physically excluded bees and butterflies from landing on the flowers. The flowers were sometimes so heavy with beetles that they actually broke off from the plant! It seems that the invasion continues.

YELLOW-HEADED ROOST

SUSAN POST

Yellow-headed blackbirds have fascinated us since our first encounter at Willow Flats in Grand Teton National Park, Wyoming. While we admired the mountain scenery, a bold flock of yellow-heads plucked insects from our car grill. The yellow-headed blackbird is robin-size and larger than its cousin the red-winged blackbird. Its Latin name, *Xanthocephalus xanthocephalus*, literally means "yellow head" and refers to the male, whose head, neck, and upper breast are bright yellow. Found in marshes near open lands, the bird is hard to see in the tall cattails and bulrushes. It is usually heard first. Its call, unlike its cousin's melodic scree, resembles a "rusty hinge" or "a guttural croak."

A small wetland in El Paso, Texas, teems with thousands of roosting yellow-headed blackbirds. A male yellow-headed blackbird calls for a mate (*bottom right*).

While we have occasionally seen these birds in Illinois and in the West, it was never more than a few individuals. So when a Texas bird-finding guide noted, "Huge numbers of yellow-headed blackbirds, almost entirely adult males, winter in west El Paso. A roost of over 6,000 is readily accessible, making for a noisy spectacle," my curiosity was piqued. A blackbird roost is a common sleeping area, usually in a wetland with dense vegetation, providing safety from predators. Preferring to avoid New Year's Eve spectacles of the human kind, we decided to ring in 2004 with a visit to El Paso. The small marsh, near a shopping mall, was not much to look at. Steep gravel mounds surrounded the marsh—a borrow pit for the mall parking lot—and

because the footing was treacherous, we could not get very close. The vegetation consisted of downed cattails surrounded by invasive salt cedar, *Phragmites*, and stunted trees. The place seemed an unlikely roost for blackbirds, or for any bird. Indeed, we were alone with no birds. Was this the right location? At 4:15 p.m., small ball-shaped flocks of blackbirds began to appear. Like schools of fish, they flew back and forth over the marsh and descended into the matted cattails with a crinkling sound, like salt being poured on aluminum foil. More and more birds came, but they were all red-winged blackbirds, and most sat on the cattail stubble. Suddenly, they flew off, en masse, even those hidden in the cattails. In the ensuing quiet, I glanced up to see thousands of yellow-headed blackbirds sitting on the nearby power lines. More and more birds decorated the lines, the males flashing their white wing spots. As if on cue, the yellow-heads began to descend, each ball of birds contracting and expanding over the small patch of matted cattails. The redwings had disappeared. Other balls of yellow-heads descended, popped down, and disappeared into the vegetation; the writhing, shrinking, expanding dark cloud of black-birds secretly slipped into slots in the matted cattails. This occurred repeatedly, the birds not "roosting" until their apparent joy of flying was exhausted for the day. The noise was loud, like rain on a metal roof. The sky turned pink with blue streaks, silhouetting the hundreds of birds still on the wires. The yellow-heads continued to descend until dark, when we reluctantly departed. The final bird species on my list for 2003 was the yellow-headed blackbird—not one, not hundreds, but thousands of individuals, filling the sky like blackbird confetti.

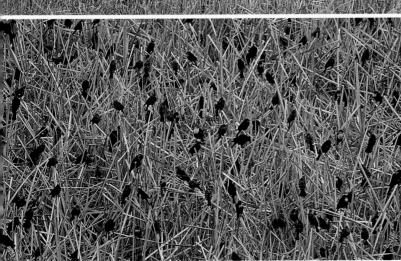

OUTBACK CAPER

SUSAN POST

We are headed into the Outback of Australia, traveling inland from Brisbane to Bowra Station, a former sheep ranch near Cunnamulla. Our journey is westward on Highway 49, an arrow-straight line of bitumen through the red dirt of the acacia-dominated mulga that lines both sides of the pavement. Mulga, a uniquely Australian habitat, is open semiarid scrubland and woods dominated by various acacia shrubs, usually gray mulga. Traffic is minimal, only the occasional truck with a dead roo (kangaroo) strapped to the top and the fast-moving, seriously long road trains—massive semitrailer trucks pulling at least three trailers. To pass the time, I count dead kangaroos along the roadside, take note of interesting signs—*Trucks Please Stop to Drop Dust before Entering Town*—and watch for free-roaming goats, another highway hazard. At times the journey seems like driving through a dusty meat locker—I will count fifty roadkill roos today. Kangaroos aren't just bouncing around the landscape, as we have had to work hard to see and photograph them, yet they seem to have an affinity for highways. White-winged choughs (crow-like birds) constantly fly up from the roadsides, but they are too common and we pay them no attention. However, a photogenic pair of Major Mitchell's cockatoos in a green acacia is cause for excitement and wakes us from our highway daze. They resemble pitchers of pink lemonade sitting in a tree. As we continue to speed inland, a white cloud, fluttering around a dead-looking shrub, catches my eye. While trying to figure out what it might be, I yell "Stop!" to Iain, our guide, and convince him to turn around and investigate. It is an emergence of white and black butterflies that I later identify as caper whites, thanks to the recently purchased massive book on Australian butterflies that resided on my iPad. Large numbers of caper-white female butterflies were emerging from their chrysalids, and eager males fluttered above them.

The Australian caper white butterfly sometimes occurs in such large numbers that their caterpillars strip the food plants. The butterfly pupates on its host plant and males emerge first, wait for females to emerge, and immediately mate with them. A Major Mitchell's cockatoo adorns an acacia *(bottom right)*.

Caper whites lay their eggs in loose clusters of a hundred or more on members of the Caper family—usually mock orange or wild caper, both species commonly found in inland Australia. When the caterpillars emerge, they are gregarious, ravenously feeding on leaves, often defoliating the entire tree or shrub where the eggs were deposited. The caterpillars pupate en masse on the skeletal remains of their host plant. The males emerge first and remain nearby to await female appearance on subsequent early mornings. What caught my attention were hundreds of males fluttering around groups of female pupae, waiting for them to emerge so they could immediately mate with them. While some courtship may be involved, most mating appeared to occur soon after female emergence. In Australia caper whites are harbingers of summer. They are also migratory, and in some years populations are so dense that thousands are killed by cars and plaster the grills of passing trucks and lie crushed on the endless stretch of bitumen, much like the roos.

BRAZILIAN CATERPILLARS

MICHAEL JEFFORDS

Under the thick green canopy, deep in a Brazilian rain forest, bright colors are a luxury, eagerly sought by those who explore. For many, color is provided by mixed flocks of shockingly patterned birds, high overhead. For the botanically inclined, epiphytic orchids cling to the massive branches of rain-forest trees, while hummingbird-pollinated red and orange tube flowers dangle from the canopy far above. For an entomologist, though, the rain forest offers a different palette. The colors I seek are just as vibrant, equally bizarre as the tropical birds, but more reminiscent of a Seurat painting. In fact, rain-forest caterpillars seem to have adopted pointillism as a color strategy aeons before French painters.

Three species of communal caterpillars—*Morpho telemachus*, a butterfly (*top left*); *Euglyphus* sp., a type of lappet moth (*bottom left*); and an unidentified species (*right*)—create distinctive, spectacular displays in a southern Brazilian rain forest.

While hiking through a southern Brazilian rain forest along the Cristalino River, eagerly seeking everything from tamanduas to toucans, I happened upon three remarkable displays of caterpillars, each uniquely notable. A tree trunk gleamed in the dark forest, literally coated with bright-orange caterpillars (perhaps tussock moth or notodontid larvae, but still undetermined) all facing downward in uniform staggered columns. From the quantity of fecal pellets around the tree, they had obviously spent the night feeding on leaves in the canopy. Their bright colors served as a warning (called aposematic coloration) to potential predators to "leave us alone." I took their advice and didn't touch them. Later, a second display, composed of hairy lappet-moth caterpillars (*Euglyphis* sp.) with persimmon-colored heads, formed a large semicircle, low on a trunk. Initially, I mistook them for a bizarre fungal growth, but closer inspection revealed that they were caterpillars. The third encounter was the most spectacular; it shone through the dark rain forest like a lighthouse beacon on a rainy night. These caterpillars were brilliant red with a filigree pattern of luscious cream. They clustered around entire branches, about head high, and obviously conveyed a "message" to all who found them—"Eating us would not be a good idea." A colleague of mine identified this red cluster as caterpillars of *Morpho telemachus*, an enormous metallic-blue butterfly of the rain-forest canopy.

What do all these creatures have in common? First, they are in some way noxious to predators and create these large displays to advertise that fact. Second, they may employ masses as a type of defense against parasitic wasps and flies. While many individuals in a cluster will end up parasitized and doomed, it is likely that some will survive unscathed and grow to adulthood. In a further bizarre twist, it was recently documented that chicks of a rain-forest bird, the cinereous mourner, may sport plumage that mimics certain species of noxious caterpillars! I have not encountered the chicks, but for me the bright caterpillar clumps decorating the dark rain-forest understory were more than enough to hold my interest.

MONARCHS IN MEXICO

SUSAN POST AND MICHAEL JEFFORDS

Since the discovery of monarch overwintering sites in 1975, seeing them in the mountains of Mexico has become a pilgrimage entomologists must undertake. We are no exception and tag along with a group from the University of Arkansas. Like the migration of monarchs, the journey is not easy. Once near the location, the only hint of the wonders above is a sign—Mariposa—and wall-side murals of monarchs. Even when we are at the site, there is still a two-mile uphill trek to where the butterflies are roosting. After their long journey, the monarchs are fickle, seldom occupying the same sites each year. So come along as we discover two entomologists' Holy Grail.

At the El Rosario Monarch Preserve in central Mexico, it is possible to stand amidst millions of monarch butterflies. The air is saturated and the ground is covered. No experience on earth is comparable to this phenomenon.

To Mexico, January 3, 2007 (MJ)—We leave Chicago for Houston with no problems. My 35 lb. camera backpack barely fits in the overhead bin. The flight is uneventful, until you consider that we flew in a noisy, hollow, aluminum tube, housing a colicky daycare center. The lady in front tried to pretend ignorance and placed her child in the seat next to her without buying the seat. Nice try.

Chincua Monarch Preserve, January 4, 2007 (SP)—After a long, exhausting, bumpy van ride, this is not what I expected. We walk through a shantytown of vendors, pay the entrance fee, and meet our guide, a wizened native lady, who has to be at least seventy. This will be her third trip of the day. She sets a brisk pace, on foot, as we head uphill; we have only our thoughts and breathless gasping sounds to accompany us. Along the way, I see monarch corpses, here and there a wing, and occasionally a live monarch slowly fluttering along the path, pausing on a flower. Forty-five minutes later, we are on top at ten thousand feet. Monarchs cluster on the oyamel fir trees. Their branches are brown and bend under butterfly weight—so many, yet I am frustrated and disappointed. The monarchs are too far away, and ropes bar me from straying any closer. Am I missing something?

El Rosario Monarch Preserve, January 5, 2007 (MJ)—I feel reverence for natural selection, for this mass spectacle. I sit amid the planet's largest puddleclub. As the day warms, I enter an orange and black ticker-tape parade set against a brilliant blue sky. I am honored to be a witness. All I can think of is a butterfly blizzard or a windy New England fall day with crisp azure skies and innumerable orange fall leaves floating everywhere in the air. Here, though, the leaves are not spent photosynthetic factories, but living, breathing butterflies, concentrated from the remote corners of North America. They congregate here to drink, nectar, rest, mate, and then return to repopulate a continent with orange confetti.

El Rosario Monarch Preserve (January 5, 2007) (SP)—Today, the trail was not crowded or noisy; even by noon, with many people in the meadow, the loudest sound was the wings—a papery falling-leaf sound. The monarch meadow evoked a magical reverence, and all we did was watch, point, and smile.

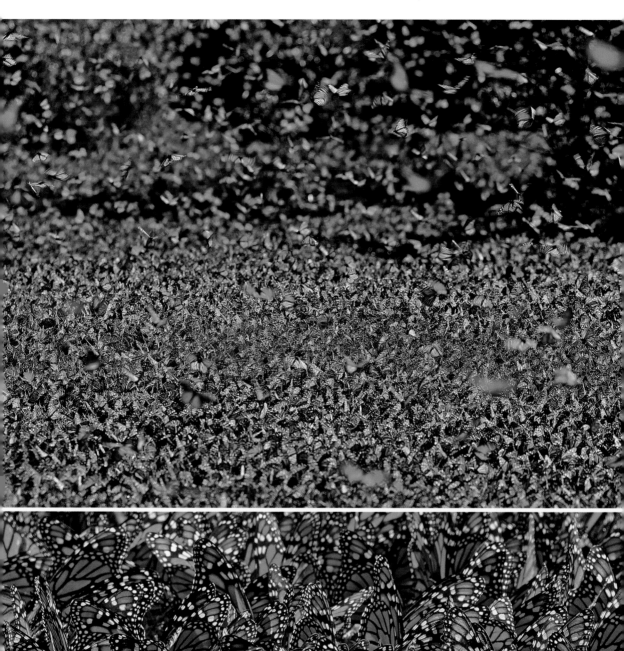

KICKING UP A STORM

SUSAN POST

No matter where we travel, one of our goals is to see large numbers of organisms. Our trip to the Sequoia groves of California was no exception. We actually managed to visit seventeen different groves. On our way back to Illinois, we passed Mono Lake, an area I was familiar with only from nature-calendar photos. I had read about multitudes of brine flies that were unique to the lake, so we had to stop. My field notes for the day stated:

A mass emergence of ephydrid flies from the saline Mono Lake triggers a California gull feeding frenzy.

Taken as a whole, Mono Lake is quite scenic, but a cursory glance from the parking lot gave the impression of desolation. The signature oddly shaped tufa deposits began in the desert, far from the lakeshore. I felt like I was walking on the bottom of a giant's goldfish bowl—the tufa deposits were the castles emerging from the expanses of six-foot-tall blooming rabbit bushes. At the lakeshore the ground was alive with thousands of dark brine flies (family Ephydridae), dancing along the edge. A few California gulls patrolled the shoreline with open mouths. What was that about? As I walked, the flies did the "wave" in front of me. Out in the lake, many strangely shaped tufa deposits reminded me of the structures mud-dauber wasps construct. The water teemed with thousands of slate-colored eared grebes, swimming and diving and gleefully eating the flies.

Mono Lake has been described as "one of the grand landscapes of the American West, an ancient lake cradled by volcanoes, glacier-carved canyons and snowy peaks." This million-year-old body of water covers sixty square miles. The water originates from eastern Sierra Nevada streams, but the lake has no outlet. As the water evaporates, it leaves behind salts and minerals, making the lake two and a half times as salty as seawater. The tufa deposits form when calcium-rich freshwater springs bubble up through the lake bottom into the alkaline water that is rich in carbonates. The compounds combine and precipitate out as limestone.

While the lake has been called a "dead sea," nothing is further from the truth. Even though only a few organisms can tolerate the lake's alkaline water, those that do thrive in amazing numbers. The lake's food chain is composed of algae, brine shrimp, and brine flies that feed on the lake's abundant algae. The flies have the ability to pump salts and other minerals out of their bodies before levels become toxic. Fly larvae live underwater and feed on algae scraped off the tufa. Laval and adult brine flies are an excellent food source for eared grebes and California gulls. Eared grebes do not breed at the lake, but they may be found during any month of the year feeding on larval flies. One birding guide described their populations "embodying the term 'abundant'" (one and a half to two million birds occur here, 30 percent of the North American population). In addition, 85 percent of California gulls begin their life at the lake and make up the second-largest California gull colony in the state. Feeding gulls start at one end of a huge mass of flies and run through them with heads down, bills open, snapping up flies. They spend their days systematically "kicking up a storm" and literally harvesting the ephydrid wind.

MITE CARPET

MICHAEL JEFFORDS

As arachnids (invertebrates with eight legs) go, the clover mite (*Bryobia praetiosa*) is a relatively large example, approaching 0.03 of an inch long, but hardly big enough to cause much concern. In fact, immense numbers of these plant-eating mites inhabit our well-fertilized suburban lawns, often residing quietly under that favorite picnic cloth or cavorting around our bare feet as we enjoy the feel of "grass between our toes." We humans are totally oblivious to their harmless presence. However, a few years ago, I had an encounter in my workplace that I could not ignore. As an entomologist, I found it spectacularly fascinating! My research laboratory, located in the verdant sward of a research park at the University of Illinois, was "invaded" in early fall by a horde of clover mites. The numbers were so great that I noticed a half-dollar-size reddish clump near my office door. With the aid of a 10x hand lens, the fuzzy edges of the "spot" resolved into hundreds, thousands, even tens of thousands of clover mites. It took only a few minutes of research to identify my new office mates, as this phenomenon of mite invasion was not unknown. However, most such invasions, if noticed at all by typical home owners, would, at best, be given a cursory glance before the small cluster would be either vacuumed into oblivion or simply swept out the door.

Thousands of clover mites invade my office in early spring.

For me, this was heaven. After assembling my super-macro camera system, I entered a world few have ever seen (or likely care to)—the swarming, swirling chaos of these red-legged arachnids. I took several photos, promptly printed them, and ventured off to share my find with my colleagues. Now it is not easy to lure busy scientists away from their computer spreadsheets or compound microscopes, but I did entice several to come down to see what I considered a once-in-a-lifetime opportunity. Suffice it to say, most—the group included a botanist, a mammalogist, and an ornithologist—were suitably underwhelmed when I pointed out the diminutive red patch on the concrete floor near my office. When I produced my photo, though, reactions picked up. The first responded with "Why are you showing me a picture of your carpet?" The second showed mild curiosity, but the third exhibited genuine concern for my mental well-being! Unfortunately, all were much less appreciative of my tiny Serengeti migration of clover mites than I was.

A search of the online encyclopedia *Wikipedia* yielded the rather mundane "Clover mites can become a nuisance in houses. They generally enter close to thick vegetation and can infiltrate houses in very large numbers." For me, even though this living carpet had vanished by the next morning, it provided an endless source of arachnid wonder, and a large photo of the event still adorns my office wall.

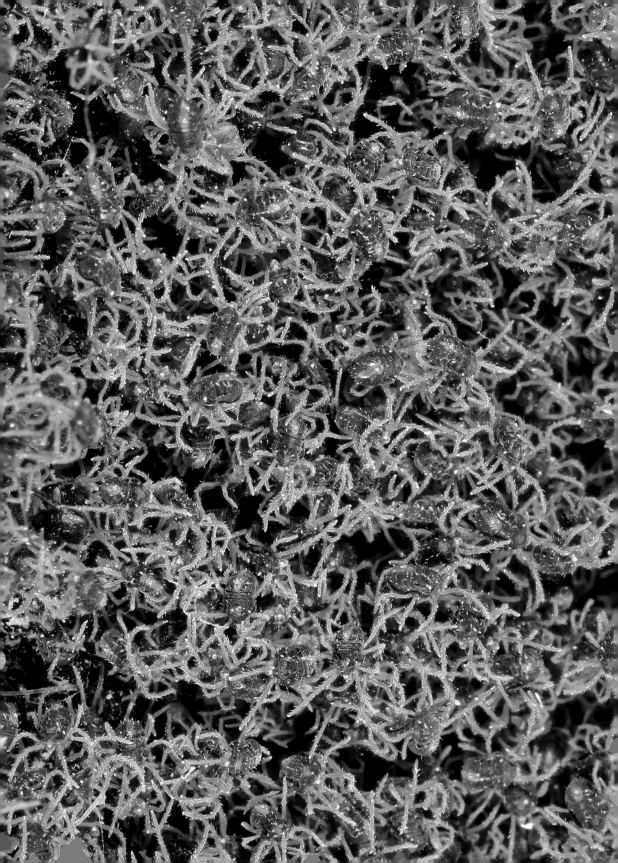

PELAGIC!

SUSAN POST

Pelagic birding (as defined by the American Bird-ing Association) is "watching the globe-trotting birds of the open ocean as they ride the air turbulence of waves." Most of my birding friends view a pelagic trip as something that must be endured to add new birds to a life list. One even commented, "The first hour I thought I was going to die, and the last four hours I wished I was dead!" All cautioned me to expect seasickness, as it is the nature of the beast. Thus, it was with trepidation that I agreed to a pelagic voyage off the coast of central Chile. We would journey to the continental shelf—an area influenced by the rich Humboldt Current and with endless birding possibilities. To prepare I not only studied seabird juvenile plumages and learned to distinguish mollymawks from prions (albatross and petrels), but also physically "trained." Rather than the typical lift-the-binoculars-to-eyes motion, I practiced balancing on a rocking Bosu ball (an exercise ball cut in half). With a four-pound weight to simulate my camera, I would balance on the hemi-sphere, look left, right, and up and down, all the while moving the weight to simulate my camera. Eventually, I found my balance and tried to encourage Michael to do the same. He replied, "It makes me seasick to just watch you!" I continued this core training until I felt ready to tackle my first pelagic, and despite the advice

Pelagic birding chaos includes Peruvian pelicans, black-browed albatross, kelp gulls, shearwaters, and several other species.

of friends, I did not medicate before the trip.

An early-morning departure from the harbor near Concon, Chile, on a small open boat with four benches—two that slid back and forth—and no hand-rails made me wonder if seasickness was the least of my worries. Two gentlemen with sea-weathered faces, fish-stained clothes, and large competent hands bus-tled about, readying for the adventure. For the first half hour, I was too afraid to stand up; I sat in place, fearful of getting sick or falling overboard, and clutched my camera to my side. Finally, I realized the boat was just a larger version of a Bosu ball, and as we rolled across the waves I was up as the first albatross appeared. As the Chilean poet laureate Pablo Neruda wrote, "[It] skims the waves with its great symphonic wings." A pattern emerged to our pelagic—up and down, side to side, with dark, sooty shearwaters angling with sweep-ing arcs, Peruvian pelicans everywhere, and masses of the ubiquitous kelp gull filling the sky.

Once at the continental shelf, we chummed—our two boatmen sat at the back with five-gallon buckets of fish parts—chopping and throwing the fleshy mat-ter into the sea with cans of fish oil. Now it was a mob scene: the high-pitched calls of shearwaters and gulls are punctuated by excited voices—"Albatross!"—as a huge bird would magically appear, but soon vanish into the rolling sea. And so I pitched, rolled, chummed, called out, pressed the shutter, and repeated—the Latin rhythm of the pelagic. As Robert Cushman Mur-phy wrote in his *Logbook for Grace*, "I now belong to a higher cult of mortals, for I have seen the albatross."

MILKWEED MASS

MICHAEL JEFFORDS

Given the media coverage and general popularity afforded the monarch butterfly, when we look at a milkweed plant, we expect to see, well, monarch caterpillars. More often than not, however, we see other insects, often great masses of them, devoutly devouring this much-hyped toxic plant. What gives? Perhaps unexpectedly, there are a large number of insects (around 450 species) that have adapted to feed on the various species of milkweeds. Many are milkweed specialists and have adaptations that allow them to thrive, despite the potent chemicals (sticky, milky sap laced with cardiac glycosides) that make milkweeds off-limits to most grazing animals.

An adult milkweed tussock moth shows off its orange and black abdomen (*above*). Large milkweed bugs (adults and nymphs) cluster on common milkweed (*right top*). An early stage mass of milkweed tussock moths will later display the typical black and orange pattern of most insects that feed on milkweeds (*right bottom*).

Two common insects often encountered on the weedy roadside milkweed (*Asclepias syriaca*) and the popular butterfly milkweed of home gardens (*A. tuberosa*) are the large milkweed bug and the milkweed tussock moth. The former (*Oncopeltus fasciatus*) is an extremely attractive black and orange insect that features the same color palette that monarchs use to warn away predators. The bugs often congregate in relatively large numbers, both nymphs and adults, on the leaves and on the seedpods. This aggregation behavior helps to emphasize that these insects are strictly off the menu

to predators. Large milkweed bugs feed on sap and on the seeds found in developing milkweed pods. Coincidentally, it is really quite easy to keep a colony of large milkweed bugs in a jar, if you are so inclined, simply by providing them quantities of milkweed seeds. This species is often used in science classes to illustrate incomplete metamorphosis and because they are large and relatively easy for students to dissect.

The second species pictured here is the milkweed tussock moth (*Euchaetius egle*), a member of the tiger moth family, noted for feeding on a number of toxic plants. A female moth will lay a large fuzzy cluster of eggs on a milkweed, and the young larvae feed en masse on the foliage, often skeletonizing the leaves—but leaving the veins, to avoid the milkweed sap. Older larvae often cut leaf veins to stop sap flow and feed in the area "downstream" from the severed vein. As larvae age, they acquire the typical black and orange coloration shared by so many milkweed-feeding insects to warn away predators. Tufts of hair appear, giving them a very distinctive appearance that has led to the common name of harlequin caterpillars. Most of the time, they disperse and feed singly or in small groups during this stage, but on occasion they can literally blanket the milkweed plant with thousands of individuals, creating quite a spectacle. The adult tiger moths have brown wings that are cryptic, but when disturbed they reveal their toxic heritage by displaying a very showy black-and-orange-banded abdomen. So next time you are searching milkweeds for the familiar monarch-butterfly caterpillars, take a closer look at the bevy of colorful insects that also call this toxic plant home.

CLOWN INVASION

SUSAN POST

One of my favorite walks in the Cache River State Natural Area in southern Illinois is behind the Henry N. Barkhausen Wetlands Center on the beginning mile of the Tunnel Hill bicycle trail. With wild landscapes on both sides, I never know what I might observe or encounter, especially in early May. My goal this day was the bridge over the Cache River, and I was hoping to be a voyeur and get a peak at prothonotary warbler courting and mating. From the bridge, I would have an elevated vantage point that would not disturb the birds. While high above the brown waters of the Cache on the aging railroad bridge, I was mesmerized by happenings of a different sort in the water below. Any notion of warblers was soon forgotten. Below me were hundreds of huge yellow smiley mouths, opening and closing in the murky waters of the Cache River. The mouths would appear, disappear, and reappear as they *glub-glubbed* water and air. It was like a scene from a clown horror movie; there were so many, I could have walked across the river on their broad backs. I assumed this was some type of bowfin gathering—an ancient carnivorous fish. One of their common names is cypress trout, referring to a favorite habitat of cypress swamps and sloughs. They live and feed near the bottom and have an interesting behavior—the males protect the

Bighead carp have invaded many midwestern streams, including the Cache River in southern Illinois. The carp entered the small river during the flood of the nearby Mississippi and Ohio rivers.

nest of fry (young fish). Several times I have witnessed a male slowly swimming and herding a black ball of tadpole-like creatures in the Cache's waters. Bowfins can breathe both water and air, so that would explain the fish coming to the surface to gulp air.

With plenty of photos, I soon headed back to the wetlands center to share my observations. A fisheries biologist happened to be there that day, and I excitedly told him what I had seen. He seemed mystified, never having heard of this behavior in bowfins. Why so many, I asked? What are they doing? He could only speculate until I showed him the photos. "Those are not bowfin," he exclaimed, "but Asian carp, probably the bighead!" No one seemed excited about this. Asian carp, a term applied to both the bighead and the silver, were brought to the southern United States from Asia for the aquaculture industry. Unfortunately, they escaped and began expanding their reach. By the mid-1990s, they became a problem in the lower Illinois River. Asian carp are aggressive, adaptable, fast-growing filter feeders—they often exceed thirty pounds. As voracious feeders, they consume 20 percent of their body weight per day in plankton, outcompeting native fish and mussels for food and habitat. While I was aware of the invasive carp in the Illinois River, their presence in the quiet Cache was not on my radar. During the spring of 2011, the area had experienced record flooding. The combined overflow of the Mississippi and Ohio Rivers had allowed the invaders—Asian carp—to find a new home. The only part of my initial observation that proved correct was that this was, indeed, a horror story, and I have not observed a bowfin in the Cache since my Asian carp discovery.

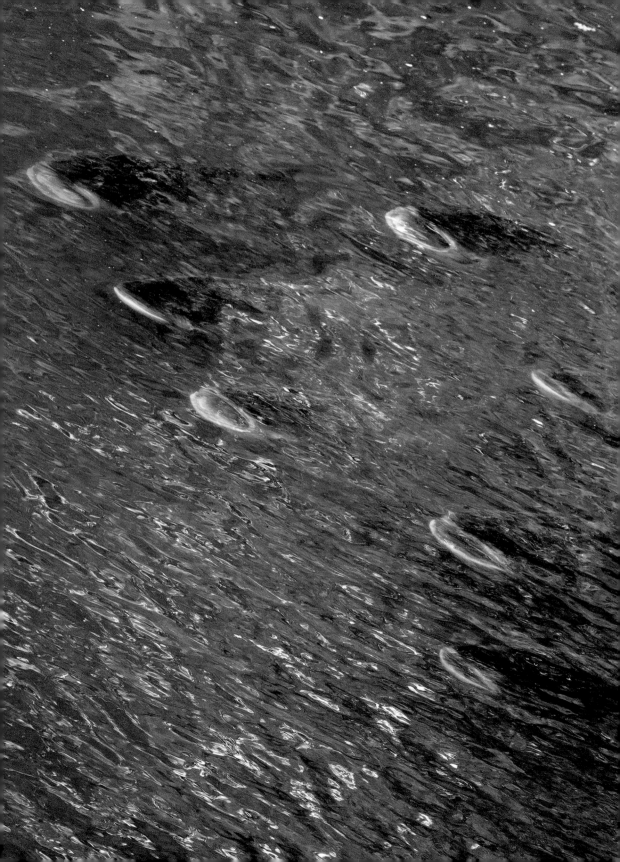

EMERGENCE!

MICHAEL JEFFORDS

I have always been interested in mass spectacles of nature. Perhaps this was instilled at an early age as I grew up on the Ohio River in southern Illinois, where I experienced the yearly summer flush of millions of mayflies. Each July at the riverside gas station where I worked, the large outdoor lights would be alive with enormous numbers of mayfly adults. In the insect order Ephemeroptera, the adults live up to their name and fly for only a day or two. They are little more than "winged gonads," as the adults have no mouthparts or digestive tracts and do not feed. Mating and egg laying in the nearby river are their only adult functions. The nymphs, however, feed voraciously and spend their aquatic life along the bottom of streams and rivers. Some cling to rocks, some burrow in the mud, others are free living and roam the river bottom searching for food (algae and diatoms), and a few are predators. Most nymphs take a year to develop, but a few species may take up to three years. Seldom noticed except during adult emergence, swarms of flying mayflies have been so large as to show up on Doppler radar! Sweeping the pavement surrounding the gas station each morning often yielded a fifty-five-gallon trash barrel full of dead mayflies.

This area also happens to be in the range of the uniquely North American phenomenon of the peri-

A mass mayfly emergence (*top*) is a yearly occurrence along many rivers. Periodical cicada adults are everywhere during an emergence. Note the red eyes on the periodical cicada face (*bottom*).

odical cicada, and I lived where Brood XXIII emerged every thirteen years. Perhaps no other insects in North America excite as much curiosity and wonder as do periodical cicadas when they make their predictable appearance every thirteen or seventeen years. Periodical cicadas are widely distributed in the eastern United States (east of the Rocky Mountains). There are seven species of periodicals—four with thirteen-year life cycles and three with seventeen-year life cycles. The species are best identified by sound, as each song is species specific. After years of living in underground tunnels, thousands of cicada nymphs emerge from the earth synchronously, as if by a predetermined signal, shed their nymphal skins to become adults, and disperse into the nearby trees and bushes. Up to forty thousand can emerge from a under a single tree! The cicadas' bizarrely long life cycle revolves around survival for the masses. When a large population of juicy cicadas suddenly appears on the scene, predators make the most of the situation, but simply cannot eat all the insects. Thus, a significant number of cicadas live to reproduce. Long-lived predators may actually remember the feast and return to the scene in subsequent years. Short-lived predators, being well fed from the cicada banquet, reproduce successfully and leave a large population to await the next year's emergence. However, "next year" does not happen for another thirteen or seventeen years, so the periodical cicada is able to outlast and escape its enemies. Perhaps Ray Bradbury described them best in his work *Dandelion Wine:* "The cicadas sang louder and yet louder. The sun did not rise, it overflowed.'" Both the mayflies and the cicadas certainly fit that mold.

THE RED AND GREEN SHOW

SUSAN POST

On the list of the world's most biologically diverse locations, Manu National Park in Amazonian Peru is number one. That is quite a statement, given the millions of species that inhabit the earth. Thousands of species congregate here, sandwiched between the Manu and Madre de Dios Rivers. We had spent a week on the pristine upper reaches of the Madre de Dios, and it was time for a daylong trip downriver to Puerto Maldonado for a flight back to Cuzco. Along the way, we had one special stop to make. At five in the morning, the full moon still rules the horizon. A common paraque breaks the silence as we start down the Madre de Dios in a narrow riverboat. The fog lies thick and heavy on the water, and Michael dozes as we motor along. I am too excited too sleep. The boatman stops at a section of the bank that looks no different from any other, but I notice a tiny trail leading up the muddy bank and into the forest. Our destination is a mile inland, and by six we are in a blind that resembles a thatched covered bridge. It is situated nearly one hundred yards from a reddish-brown clay bank along an old streambed, topped with dense rain-forest vegetation. Inside the blind, silence rules, but outside all is a chorus of squawks and cheeps. Blue-headed parrots nervously fly back and forth in small, noisy flocks in front of me. They never land. Barely visible in the morning mist, the distant clay bank has a few yellow-headed parrots. By seven, three or four red and green macaws have

Red and green macaws gather at a clay lick along the Madre de Dios River in Amazonian Peru.

flown in. They are mantled with red and have long, sharp tails and bright-blue wings. For the next hour, more macaws come in, fly away, and return a short time later. They preen and bill fight, usually in pairs or threesomes, clinging to the rain-forest foliage. Like the morning mist, the silence has lifted. The sound is distant, but deafening. At one point, at least two dozen macaws appear, followed by more. Though still in the forest above the clay lick, they move lower and are at the edge. Momentarily, they all arise, en masse, and fly up the old stream channel, disappearing into the haze. Is that it? Only a few minutes later, they return to the trees, and by a quarter past eight a single macaw finally comes down to the clay. Others follow, and soon it is a kaleidoscope of red, green, and blue—colorful, noisy, and chaotic against the tan clay. Macaws congregate at these special places to ingest the alkaline clay. It is thought that macaws consume certain seeds that contain toxins and caustic materials; the clay helps neutralize these compounds in the bird's digestive system. By half past eight, most have left, and the clay is again silent. Even the large flocks of blue-headed parrots have left as the day heats up. It is time for us to go, and by nine fifteen we are back on the boat, headed downstream. All I hear is the whine of the motor, the chop and splash of the brown tropical water, and the wind in my ears. The natural world has gone silent.

Over the next nine hours, I have time to ponder my experience. While I might have thought of a clay lick's color as just reddish tan, my experience on the Madre de Dios illustrated that merely by adding accents of lime green, yellow, red, and sky blue, the lick came alive with a show I will never forget.

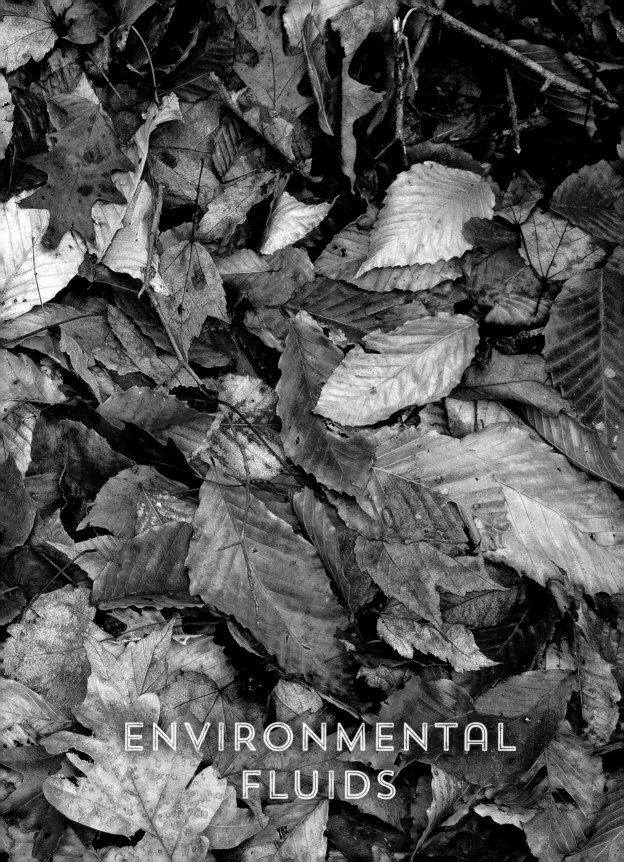

ENVIRONMENTAL
FLUIDS

The term *environmental fluids* inevitably conjures up images of water. In landscapes across the world, water is a dominant influence, either by its overriding presence (bogs, swamps, marshes, and rivers) or by its scarcity (deserts, polar regions, or landscapes with high relief). On a December hike in the Cumberland Mountains of East Tennessee, we witnessed the power of this universal fluid and came to better understand the simple phrase *water erosion*. The East Tennessee landscape is moist, even rife with water. We experienced a mixed conifer and deciduous forest while hiking in the rain—actually, after several days of rain. The forest floor was a carpet of soaked vegetable leather leaves, defined by their individual shapes. Red oak, white oak, and chestnut oak leaves—some pointed, others bristly, others rounded—interwoven into a flat, chaotic matrix. It was like looking too closely at a sheet of handmade paper. Our favorites, though, were the beech leaves. They formed a carpet reminiscent of the back of a long-dead giant reptile, the moist scales just ready to slough off the large decaying carcass.

Wet leaves carpet the floor of a Tennessee woodlands. (*previous page*). A female ouzel bird shakes water from her wings after a dip in the torrent (*left*).

Paradoxically, water erosion is often explained in geological textbooks in an uninteresting, theoretical way—such tedium. That day, however, we experienced it firsthand. Our hike led us through epic sandstone overhangs or undercuts—at least two dozen of them. The driving rain and our own disgruntled attitudes made the hike unpleasant. Viewing the dynamic nature of erosion, however, was well worth the effort. Water was everywhere, falling from a hundred new waterfalls, moving sand from here to there, trickling through the stone to nowhere. Even the normally dry deep overhangs were soaked. Everywhere we touched, the sandstone was moist, and each cemented grain of sand surrounded by the universal solvent, water. Drip, flow, cascade—the process was unmistakable. We hiked in front of waterfalls, misted by their spray; behind waterfalls, annoyed by the drip-dripping of moist rock; and through waterfalls, drenched and pelted by droplets hitting us with the force of BB pellets. Multiply all this activity caused by a single rain event, by days, by months, by seasons, by years, by decades, by centuries, by millennia, by epochs, and suddenly, all was clear. We understood and appreciated water as a primary mover of land.

Conversely, on a trip to Arches National Park, both abundant water and the lack of it played a significant role in our adventures. Southern Utah is basically a desert, receiving less than ten inches of rain each year. Yet we were awakened as we floated on a river of silt rushing under our tent, triggered by a distant flash flood. Later that day, we hiked to near dehydration on a "short, easy trail" when we failed to bring along enough water. In Zion National Park, we explored a spectacular run-and-riffle river and had a memorable encounter with a trio of water ouzel birds (also called American dippers). One savvy adult dipped, bobbed, and surfed its way among a limitless smorgasbord of aquatic macroinvertebrates that lived in the boulder-strewn stream. Her two fledglings watched and played at being adults—they had the dipping and bobbing down, but seemed reluctant to enter the water. When they did, they bounced helplessly in the swift current. Fortunately, mother ouzel emerged from each immersion in the torrent with a tasty morsel for her offspring.

In this short section devoted to environmental fluids, water is obviously a central player in each essay, but in less familiar ways than described above. Water serves as a carrier for various semisolids, exhibits unique configurations associated with changing weather patterns, and is the solvent for necessary nutrients required by a variety of animals.

MAY 28, 2011

SUSAN POST

When I tell people I live in Illinois, they look at me with skepticism and sometimes sympathy. Wouldn't California or Colorado be a better place for a biologist? At least there are things to see in those states. However, I am proud of my state and will pit a good day in Illinois against one anywhere else in the Union. May 28, 2011, was just one of many memorable days I have experienced in Illinois. Late May in southern Illinois is sublime: the trees have leafed out, and it is warm. Michael and I visited several sites that day, looking for butterflies. While the zebra, tiger, and pipevine swallowtails had been on the wing since April, our quest was for rarer species, like the Ozark checkerspot. A nice feature of southern Illinois is that many different habitats can be visited in a day—nothing is too far apart, and traffic is minimal. We drove past Cave Creek Glade Nature Preserve, where the pale-purple coneflowers bloomed on top and on the slope—bright-pink color ran down the green hillsides like topping on a strawberry sundae. Our first stop, however, was Simpson Creek Barrens, tucked into the Shawnee Hills. We parked at a split-rail fence and contemplated if we really wanted to explore the area, as ticks were using the fence as a highway, waiting for any mammalian visitors. We donned rubber boots, applied copious amounts of repellent, and began our

An Ozark checkerspot butterfly feeds on exudates from the shell of a recently deceased Carolina box turtle (*top*). Perhaps the largest butterfly puddleclub we have seen (including in the tropics) gathered on the banks of Barren Creek in southern Illinois (*bottom*). What you see here is only a small part of the total.

foray into the grassy glade. Michael found a hognose snake, and we both photographed it while it did its best "I am dead" routine. I spied a spider's web laced with a pair of butterfly wings. My first impression was gray hairstreak, but the wings just didn't look right. Michael agreed, and we put the treasure in a glassine envelope to share with our butterfly colleague Jim Wiker. We continued our survey of the barrens, admiring blooming Indian pink, a newly emerged goatweed butterfly, and numerous great spangled fritillaries. I soon located what we had come to see. Hidden in the matted grasses rested an Ozark checkerspot, its orange antennae aglow, as if they had captured and were reradiating the warm sunshine. As we watched, it probed a dead box-turtle shell, imbibing the last blobs of nutrient goo from the deceased reptile. With fading repellent, we moved to a new area and by late afternoon found ourselves on a trail to Barren Creek. While stumbling over roots and brushy tangles, I noticed quite a few swallowtail butterflies in the air. At the creek bank, we were stunned by one of the largest butterfly puddleclubs we had ever witnessed. More than a hundred tiger and spicebush swallowtail males gathered on the moist ground, and more kept coming. Any disturbance sent them flying, swirling, only to return and land again. A puddleclub is a group of bachelor male butterflies gathered at a moist spot that contains salts, minerals, or some other compound they find attractive. What a day!

Once home, I contacted Jim Wiker and gave him the envelope with the tiny gray wings. A week later I got a call. "Sue, you are not going to believe this, but those are the wings of a northern hairstreak. I've only seen two of those in my forty years of collecting. You guys had a really good butterfly day." Yes, Jim, I would have to agree.

HERPTILE SIPPING

MICHAEL JEFFORDS

Tropical rivers are magical places. The diversity of life there is staggering, both in the number of species and in the complex interactions that have evolved among unrelated species. The Madre de Dios River that runs along the southern edge of Manu National Park in Peru is a remarkably untouched stream in its upper reaches. Lined with primary rain forest, the river passes through some of the most biologically diverse landscapes on Earth. A day spent exploring here yields a lifetime's worth of experiences, including many sightings of the yellow-spotted Amazon River turtle. Equally amazing is the Cuiaba River that flows through the Pantanal wetlands of Mato Grosso in southwestern Brazil. Here vast numbers of birds and reptiles can literally carpet the quiet, still waters of the bordering wetlands. The Jacaré caiman population of the Pantanal approaches ten million individuals and is perhaps the largest concentration of crocodilians on the planet! On separate explorations in both of these ecosystems—jetting around the Cuiaba River in a speedboat looking for jaguars and quietly floating down the Madre de Dios River in a longboat seeking many species of birds—I witnessed interactions between the aforementioned reptiles and my favorite group of animals, the butterflies.

As an entomologist, I am familiar with the behavior

A Jacaré caiman hosts heliconia butterflies taking a sip (*top*) in the Pantanal. Yellow-spotted river turtles bask with a lone swallowtail on the Madre de Dios River (*bottom*).

of various species of butterflies that often land on my sweaty skin to imbibe the salt-laden moisture. While butterflies feed on nectar (rich in sugars), most species also require micronutrients (salts, nitrogen) that must be garnered from other resources. In the tropics, butterflies are especially creative in where they find these various nutrients. This phenomenon is often featured on various nature programs about the tropics. Most crocodilians and turtles have glands located on the head to rid the body of excess salt (especially those species that frequent saltwater). While both the Jacare caiman and the river turtle are freshwater species, they must also rid their bodies of these superfluous chemicals. A review of the scientific literature confirms this, and I also witnessed the phenomenon of tropical butterflies—long-wings and swallowtails—surreptitiously sipping secretions from both the caiman and the river turtles. In fact, there were often dozens of butterflies flitting about and regularly landing on the heads of both species of basking reptiles. I could see they were feeding as their uncoiled mouthparts were inserted into the nostrils and eyes of the obliging hosts. Observing this was one thing; trying to photograph it was another matter. Attempting to convey to our boatmen (no English for them, no Spanish for me) the need to stop in the middle of the river for a mere butterfly sitting on a turtle or caiman when often dozens of macaws were flying overhead or even a jaguar was prowling the shoreline proved to be as big a challenge as any I encountered on the exploration. Eventually, the boatman and I came to a nonverbal "agreement," as he recognized that I was not the typical "show me only the big stuff" kind of guy.

FRAGRANT FIELDS

MICHAEL JEFFORDS

Being a biologist can have its hazards. Giving presentations to the general public might appear to be a rather tame type of fieldwork, but not always. I had innocently accepted an offer to be the banquet speaker for a southern Illinois organization. The meeting was held at a local restaurant—called Fragrant Fields—in the small town of Dongola. The restaurant was a converted mule barn, quite charming, and the evening's event was well attended. The dining area was adjacent to a short raised platform—obviously the area that had held the mule stalls—that was covered with pea gravel and dotted with a few potted plants. My host suggested that I set up my screen in the pea gravel, as this would not disrupt the diners and would provide a good presentation venue. After erecting the screen, I noticed that my shoes were coated with a thin layer of pea gravel. Not really thinking anything about it, I swiped the gravel off with my hand. At that moment, someone asked me a question, and I inadvertently swiped my hand across my nose and face. Wow, what an incredible sensation that produced! The material binding the gravel to my shoes turned out to be cat feces, and I had managed to fill both of my nostrils with this odoriferous mass. My then brown beard also managed to entrap a significant load of cat crap. It seems the mule barn and restaurant still had some openings to the out-of-doors. Evidently, the local cat community had been entering the building to use the gravel pad as a convenient litter box, and judging by the number of now obvious piles, they had been doing it for some time. So much for restaurant inspectors! There I stood, in dress pants and a white shirt, at the front of a room full of happy diners, with my face and hands full of feline semisolids. This required quick thinking as to how to extricate myself from this dilemma, but nothing in my scientific training had prepared me for this. I just stood there. Fortunately, the room was dim, and no one was paying much attention to me at this point, so I quickly exited "stage left" into the men's washroom. It took a serious amount of time to *decrapify* myself, and I left considerable evidence in the sink. Although the next patron may have wondered why the sink was laced with pea gravel, the odor could have been explained by, well… you know. It took some time before I managed to remove the last vestiges of "feline essence" from my hands, nostrils, beard, and shoes. Miraculously, my shirt had remained a "feces-free zone." I would have had a hard time explaining why I had decided to give the keynote talk in my T-shirt. Fortunately, the remainder of the evening proceeded quite normally, the talk went well, and I carefully removed the screen from the gravel before disassembling it for the trip home. Not too long after that experience, I learned that Fragrant Fields had closed it doors.

From a mule barn to the Fragrant Fields restaurant to the Barn Shoppe, the building provided few clues to the drama that unfolded one evening. Today, the barn is gone, replaced by a Caribbean restaurant. Fortunately, it is housed in a cat-proof pole building, and the food is pretty good.

SPIDERS, STRIDERS, HEXAGONS & HALOS

MICHAEL JEFFORDS

In high school chemistry class I was taught about the characteristics of water. Predictably, I failed to appreciate fully this unique liquid. Water is the only substance on Earth that occurs naturally in all three phases—solid, liquid, and gas. It dissolves more substances than any other common liquid, has the highest surface tension, and is only slightly compressible. But perhaps its most important feature is that it is necessary for all life on Earth. Water is the "universal environmental fluid." Over the decades, I have had occasion to observe, and now fully appreciate, many of the properties of water, each displayed in an unusual way. A late-winter visit to Mermet Lake Conservation Area in southern Illinois showcased hundreds of ice spiders in the partially frozen lake—formed by water flowing from under the ice into snow that covered the ice sheet. A central hole had allowed water to infiltrate the snow layer, giving the "spider" its characteristic shape. Remarkably, the center of each ice spider contained a dead fish! I can only speculate that the freeze caused a late-season fish kill, and perhaps the dead fish formed a warm nucleus around which the ice melted to create the initial opening in the ice.

A second phenomenon occurred during a freak winter snowstorm in the swamps of southern Illinois. Most winters in this area are mild, but in the past decade, capricious weather patterns have been the norm, and short-lived, intense blizzards often occur. When such a storm was predicted, a colleague and I headed south to witness the event. Our last stop of the day was at Heron Pond State Natural Area, just as the sun was setting. While incredibly beautiful, the scene was quite puzzling—it appeared as if the snow had accumulated only around the bases of buttressed bald cypress. Also, there was a remarkable pattern of multisize hexagons, each with a dark center, present in the partially frozen swamp. After a little thought, I deduced that the snow halos around the trees were likely caused when the blizzard deposited snow on the tree trunks. Later in the day, as the temperature warmed, the snow on the surface of the pond turned to slush, and the snow simply slid down the trunks to form the halos. The ice hexagons, however, have not been so easy to explain. Scientists know that in nature, hexagons are an efficient shape—note honeycombs and the hexagonal patterns of interlocking basaltic columns associated with volcanic landforms. In these instances, the hexagons are uniformly sized, but those in the swamp were of varying sizes. That swamp mystery remains unsolved.

A final observation involved the surface tension of water. Any number of insects and their relatives (Arthropods) use this unique feature in their daily lives. Perhaps the most familiar example involves water striders (family Gerridae). These insects rely on the surface tension of water to hunt. The geometry of their bodies and legs helps to distribute their weight across a wide area, and tiny hydrophobic hairs on their feet repel water and act as "skates." These skates create tiny dimples in the water's surface.

While all these observations have an explanation grounded in pure chemistry or physics, as Loren Eiseley so eloquently stated, "If there is magic on this planet, it is contained in water."

A cypress swamp after a blizzard exhibits both snow halos and surface hexagons (*top*). An ice spider contains its attendant dead fish (*bottom left*). A group of water striders congregate on the still waters of a large lake, buoyed by the surface tension (*bottom right*).

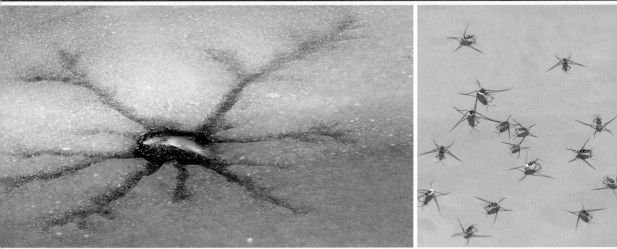

HIDDEN WORLDS

We had just completed participating in our second "Biodiversity Blitz"—a twenty-four-hour inventory of all the species in a given locality—and came away with unexpected insights into the subtleties of nature and natural history. The blitz took place in the Lake Calumet region of Cook County, Illinois, and involved a variety of scientists from many disciplines—each passionately engaged in documenting their own corner of the biotic world. A robust fisheries biologist, clad in damp trousers and a white T-shirt that was slit at every joint for comfort, drove into the parking lot near the blitz headquarters tent pulling a large johnboat. He immediately began to showcase the fish he had captured from nearby Wolf Lake. Each small and large species produced the appropriate *oohs* and *aahs* from the crowd. From the tent emerged another type of scientist, an amateur protozoologist, who ambled over to the boat. He was carrying a small plastic shoe box. His interest in the large fish was peripheral, as his real desire was to sample the surface of the fish for ciliates (a group of one-celled animals called protozoa). Obligingly, two largemouth bass were offered up. "Do you want me to clip a fin or maybe pluck a few scales?" the biologist asked. "Nope," the protozoologist said. "Just the slime, please." He simply took a microscope slide and scraped the fish's side and then happily scurried back to the tent with a few microliters of potentially wonder-laden fish goo in a small petri dish. We were smitten, by both the singularity of purpose and the diversity of perspectives these individuals brought to the same enterprise.

By the end of the blitz, both scientists, though working with widely divergent methodologies, had generated equally impressive species lists that contributed to the recorded diversity of the site. This little drama struck a chord of familiarity as we realized that this is the way naturalists approach depicting the natural world. Each has their own perspective and extracts individualistic observations from the landscape. Ten people can go to the same site at the same time and return with very different collections of things that reflect their own personal biases, experiences, and interests. For this group of essays on hidden worlds, we collectively and individually seek to extract the essence or heart of an ecosystem. As biologists, it is essential to look for the hidden elements that make landscapes unique—or, perhaps more succinctly, to seek "just the slime, please."

Sue intently stalks butterflies in a limestone glade in southern Illinois (*previous page*). A pair of largemouth bass is displayed at a northern Illinois Biodiversity Blitz.(*left*).

ATTRACT OR REPULSE?

MICHAEL JEFFORDS

My first class in entomology in graduate school at the University of Illinois was taught by a world-renowned chemical ecologist, Dr. Robert Metcalf. His work was central to the eventual ban of DDT in the 1970s. He was also instrumental in the development of a widely used group of insecticides (for example, Sevin), but his first love was the chemical ecology of insects. This branch of science investigates how chemicals released by insects affect the behavior of other organisms. His favorite story was about the time he was working on the synthesis of methyl eugenol, a volatile component found in many essential oils. After a long day of chemical bench work in an un-air-conditioned lab, which included constantly dabbing sweat from his forehead with a handkerchief, Dr. Metcalf returned home and went to bed. Before retiring, he had laid his handkerchief on the screened windowsill of his bedroom. When he awoke the next morning, he gleefully noted that the screen was literally covered with fruit flies! Thousands of them whirled, buzzed, and walked about. What was up? It did not take him long to determine that the culprit was the compound on his handkerchief, methyl eugenol. It had an irresistible attraction for this group of small flies.

Over the semester, I learned much from Dr. Metcalf, including a bevy of new terms—*pheromone, kairomone,* and *allomone*—that were variously attached to the limitless complex chemical compounds he ceaselessly sketched on the blackboard. Pheromones are most often associated with mating, as they function to attract individuals of the opposite sex. Kairomones are chemicals released by animals that do not benefit the releaser but attract other organisms, such as predators and parasites—almost like chemical eavesdropping. Allomones, however, are usually noxious compounds released by both plants and animals that serve to protect the emitter from predators.

All of this information resided in my brain over the years, but with the exception of pheromones, they had little professional relevance, as they were never part of my research. However, a few years ago, it all came flooding back when I encountered a very large wheel bug (Hemiptera: family Reduviidae) hopelessly entangled in the web of a garden spider. I noticed it because the insect's bright-red scent sacs extruded from the end of the abdomen. Most Hemiptera have glands for chemical defense, but they are located elsewhere on the body. These release noxious chemicals (allomones) that help the bugs deter predators. On this occasion, though, it did not seem to be working so well, as the wheel bug was definitely not going anywhere, whether the spider chose to eat it or not! The most interesting aspect, however, came from a closer inspection of the doomed insect. It was covered with tiny fruit flies, running up and down its silk-shrouded body, apparently attracted to—my memory soon dredged up those long-ago lectures—methyl eugenol. It appears the wheel bug's repellent allomone likely contained this attractant, discovered long ago by Dr. Metcalf. It seems the wheel bug's allomone is also a kairomone for fruit flies! Is this just a chemical ecology "accident"? We may never know, but I am sure Dr. Metcalf would have been intrigued by the entire scenario.

A wheel bug has become encased in a garden spider's silken web. Note the red scent glands and the many tiny fruit flies on its body (*top*). A wheel bug is a vicious predator on other insects (*bottom*).

SUPERNORMAL STIMULUS

MICHAEL JEFFORDS

Sue and I are decoy collectors, and over the years we have occasionally come across "magnum decoys." These are larger-than-life versions of ducks and geese that hunters thought would be more likely to attract their quarry, rather than the regular-size fake birds. A little investigation as to why this should be a factor led me to the phenomenon of a "supernormal

Supersunflowers are created by a crab spider; note the prey item (honeybee) (*top*). A normal-size sunflower with a crab spider still awaits its prey (*bottom*).

stimulus." Investigated by such notable scientists as Konrad Lorenz and Niko Tingergen (both studying birds), their research caused them to conclude that "a supernormal stimulus is an exaggerated version of a stimulus that provokes an innate response that has developed in an organism over time." Simply put, animals, including humans, can be affected more strongly by "exaggerations." The "junk food" industry thrives on this tendency, as do certain industries associated with human sexuality. It's almost like "highjacking" previously evolved behaviors for the benefit of those producing the supernormal stimulus.

We know that supernormal stimuli are a reality, especially those created by humans (for example, advertising) and that they often elicit strong responses. But do they evolve in nature? I encountered what could be construed as a spectacular example during a visit late one summer to Goose Lake Prairie State Park in north-central Illinois. The park—an expansive, thousand-acre tallgrass prairie—in late summer is a golden sea of sunflowers, stretching to the horizon. During this visit, I noticed that several of the flowers appeared unnaturally large, often six or seven times bigger than their nearby companions. My curiosity piqued, I investigated six of these "giant flowers" and found that individual crab spiders, using their silk, had woven the blooms together into large, flat platforms. These immense superflowers were highly visible in the surrounding expanse of yellow. In addition, six out of six crab spiders responsible for the megaflower clusters had successfully ambushed a prey item. Their prey ranged from honeybees and bumble bees to small moths. Hmmm… Evolved supernormal stimuli? I looked for traditional crab spider haunts and found several spiders residing in single blossoms and noted that none of these had any recently captured prey. While certainly not a "definitive" study, the superflowers appeared to be more attractive than single flowers, and this translated into enhanced crab-spider-hunting success.

How could this example of a supernormal stimulus have evolved? Perhaps certain spiders produce more silk than others, and this silk (extruded when spiders are disturbed or move about) served to bind nearby flowers into the observed "superflowers"—initially, a happy and fortuitous accident. If these crab spiders were more successful at food gathering and ultimately achieved superior reproductive prowess, who knows what evolution could bring about, given a few thousand spider generations? I have observed this phenomenon on only a single occasion and have been unable to investigate further, so the proverbial "scientific jury" is still deliberating.

THE EGRET AND THE FLY

MICHAEL JEFFORDS

While the title may seem like a children's fable, it certainly is not. This is a tale of woe (for the bird) and opportunity (for the fly). Called the most elegant of the egrets, the snowy is a denizen of coastlines, but can be found inland from the north-central United States all the way south to Patagonia in southern South America. Its pristine white plumage is distinctive, along with its black legs and bright-yellow feet. Stalking prey along the margins of wetlands and along the seashore, snowy egrets are the epitome of grace and beauty.

A snowy egret displays its parasitic hippoboscid fly, peering out from the safety of its feathery world.

I was photographing the individual shown here at Ding Darling National Wildlife Refuge in Florida early one May when I noticed a dark spot on its immaculate plumage. As an entomologist, I thought perhaps this could be a glimpse into the inner darker world of beautiful birds and approached as close as I dared. As I peered through the viewfinder, the dark spot materialized into a distinctive shape that I recognized from my entomology textbooks. A hippoboscid fly was alternately peering out and retreating back into the safety of the head plumage of this stately bird! One seldom witnesses the dark realities of nature, especially the intimate, cryptic, and often unknown world of parasitism. This was a fascinating discovery; I had never seen this family of flies before, and I was eager to learn more about them. The snowy egret would not tolerate my close approach and flew away, leaving me to ponder this unique sighting.

Afterward, I did a little research to understand the world beneath those snowy-white feathers. Hippoboscids are obligate parasites of mammals and birds, meaning they are dependent on their hosts for survival. Their common names are *keds* or *louse flies*. Some species are wingless, and others can fly. The egret's fly seemed to be of the winged variety, and I noted that some louse flies can actually transmit disease within a bird community. A check of the taxonomy of the family revealed that five genera start with the prefix *Ornitho* (relating to birds), so I assumed that my little avian hitchhiker belonged to one of those. Louse flies feed solely on the blood of their hosts, and of the more than two hundred known species, 75 percent feed on birds.

I later learned that many species not in the *Ornitho* genera parasitize birds, so my earlier assumption proved to be premature. Only by having the fly in hand could I have correctly identified it. Some louse flies are so specific that they live and feed on only a single species of bird—for example, there is a frigate-bird louse fly and a booby louse fly. Fortunately, given the choice between their normal host and humans, they choose to stay put. They have been induced to bite humans and leave an itchy bite, but cannot survive or reproduce on us. This is likely a comforting thought for ranchers who keep sheep (sheep keds) and for equestrians (horse louse flies), who are sure to scare up a fly or two during shearing or grooming. The reason I was able to identify this fly to family was because of its shape—flattened dorsoventrally (from top to bottom)—which allows them to slip easily beneath the elegant, showy plumage of their chosen host. Since that encounter, I never look at a bird without wondering what is hidden from view by its lovely feathers.

YARD BOT

SUSAN POST

One day in May, as I observed our multitude of gray squirrels during their daily backyard feeding, I noticed a very "special" individual. It had a marble-size lump on its jaw, giving it an asymmetrical chipmunk cheek. I thought it odd, but then promptly forgot about it. A month later, as Michael and I were watering our geraniums, we noticed a large beelike insect (a fly) nestled in a bloom. He commented, "Wow, look at that bee mimic," and I said, "It's so good, we'd better get a photo." Michael dutifully got his camera, and as he got closer he discovered the fly had no mouthparts. He became quite excited, calling me over to take another look. "This is an insect family you just don't see, especially the adults!" he exclaimed. What we had discovered was a member of the family Oestridae—a tree-squirrel bot fly, *Cuterebra emasculator*—a parasite of tree squirrels and chipmunks throughout eastern North America.

This gray squirrel is infested with a squirrel bot fly. Note the distinctive warble on its cheek (*top*). A squirrel bot fly (*bottom*) found a few weeks later is not from the pictured squirrel, but from last year's crop.

A parasite is an organism that lives in another species, usually without killing it. Though this fly was on a flower, it was not visiting the plant for food—adult bot flies do not have mouthparts, so they cannot eat. They also do not bite or sting. The female lays her off-white, oblong eggs not directly on a squirrel host, but on twigs, branches, or vegetation in the habitat of the hosts. The legless larva is called a bot. The developing first-stage (instar) bot remains within the egg until the body heat from a nearby squirrel stimulates it to hatch. The first instar is the infective stage, off-white in color and encircled with bands of black spines. When the bot contacts a squirrel, it will enter through one of the animal's bodily openings (mouth, nostrils, or anus) or a wound or burrow into the skin. Once inside the squirrel, the bot travels through the body and settles underneath the host's hide, molts to the second instar, and creates an opening to the exterior. This hole allows air to enter and provides a route for the elimination of excrement. The presence of the bot stimulates a response in the squirrel's tissues—a pocket (called a warble) forms that encloses the larva. I had observed a marble-size lump on the jaw of our special backyard squirrel that was a warble.

Squirrel bots have three larval instars, and only one bot is found per warble. It feeds on lymph fluid, cellular debris, and leukocytes (white blood cells) of the host. Development usually takes three to four weeks. The mature larva emerges from the host by backing out of the exterior hole, dropping to the ground, and burrowing into the soil, where it pupates. This species of bot fly overwinters as a pupa buried in the soil—for anywhere from eight to ten months—and emerges in early summer, seeking a mate. Thus, the adult bot fly we found had developed in an individual from last year's squirrel population, meaning our local squirrels had an ongoing bot fly infestation. Squirrel bots are native parasites, and except when the infestation is high—many warbles per squirrel—there is usually no detrimental effect. However, when they are in a "sensitive area"—note the name *emasculator*—well, you get the picture. The empty warbles usually heal within a week.

What happened to our bot fly? After many photos, the fly was given to an eager entomology graduate student for his collection. After all, it is a rare family, and our squirrels were certainly grateful to be rid of it!

REALITY VERSUS BOOK LEARNIN'

MICHAEL JEFFORDS

Sir Arthur Conan Doyle's legendary protagonist, Sherlock Holmes, was a master at seeing bits and pieces of a story and putting them together to explain what happened. In graduate school, students are trained to look at the world in somewhat the same fashion. They develop a working hypothesis and then ask questions that can hopefully be explained

Planidia larvae of a blister beetle clustered on a leaf (*left*) and on the flowers of bishop's cap (*top right*). The adult that likely deposited these larva is quite large, flightless, and colored to let predators know that she is highly unpalatable (*bottom right*).

after performing elegant experiments and accumulating observations that ultimately lead to unambiguous answers. If this happens, the reward is the coveted "sheepskin," followed by a lifetime of repeating the process. Such is academe. But what about the knowledge we acquire that is not immediately applicable to our research or that we do not soon put to practical use?

This was driven home to me several years ago during a visit to Great Smoky Mountains National Park (GSMNP) in Tennessee. While hiking a steep mountain trail photographing early-spring wildflowers, I happened upon a bishop's cap with its tiny white flowers. Surrounded by acres of great white trilliums, I might not have noticed the diminutive flowers, except that one plant was coated with tiny highly active insects. On close inspection, I determined them to be... well, I really had no idea! So much for that Ph.D. in entomology—I could not even place them in the correct insect order. Oh, the horror of not knowing something so simple! Even though I completed the trip and came home with thousands

of images, I constantly thought about that group of unidentified insects in my photos. What could they be?

Later that same season, in central Illinois, I was photographing a large blister beetle (*Meloe*), and something clicked in my brain that took me back to basic entomology. I instantly knew what the images from the Smoky Mountains had to be. For students studying entomology, a class in taxonomy is a requirement. Students are expected to learn all the insect orders (thirty-one or thirty-two) and a significant number of families (there are upwards of five hundred in the United States alone). Beetles in the family Meloidae (called blister beetles because their bodies contain a noxious compound—cantharidin—that causes blistering on human skin) have a rather bizarre life cycle. A female blister beetle will deposit a large clutch of eggs, often on a plant. When the eggs hatch, the larvae are not the slug-like grubs of many beetle families, but highly active roamers that seek out other insects. If a bee visits a flower where these larvae (called planidia) are present, they will crawl onto her body, often en masse. They use the bee to hitch a ride back to her nest (called phoresy), where they crawl off and become parasites on the helpless bee larvae. Bumble bees and other ground-nesting bees are their favorite "rides." They will even eat the ball of pollen the bees have provided for each bee larva! What I had seen on the bishop's cap was a group of blister-beetle planidia larvae waiting for a bee so they could hop on. I even dredged up a mental image of the page in my taxonomy textbook where this activity was described. "Awesome," I thought, but I was disappointed that I had failed to immediately recognize what was happening. Had I remembered, I could have waited around to photograph the spectacle of planidia climbing aboard an unsuspecting bee. Oh, well, maybe next time!

ACORN WEEVIL

MICHAEL JEFFORDS

There is a bit of "squirrel" in many of us. We cannot resist the urge to pick up that perfect little acorn fallen from a stately oak. These acorns often end up in a bowl on the kitchen counter with keys and other junk and are soon forgotten. There they stay, out of mind, until one day we reach for the keys and encounter a soft, squishy maggot-like creature in the bottom of the bowl. If you are a curious sort, then the question is "Where did this creature come from?" A close inspection of the larva will reveal a dark-brown head on a grub-like body; maggots can be ruled out, as they have only a pair of hooks where the head should be. The only items in the bowl—other than keys, coins, and the odd ballpoint pin—are those forgotten acorns. A check of the acorns reveals the source of our visitor. At least one will have a small, round hole. What's up? Congratulations, for you have encountered a charming insect called an acorn weevil.

An adult acorn weevil is large by weevil standards at three-eighths of an inch (*top*). A female acorn weevil assesses the suitability of a developing acorn for her eggs (*bottom*). An acorn weevil larva just emerged from its acorn; note the emergence hole (*inset*).

Another question asked by the inquisitive mind might be "How did the creature get into the acorn, as there were no holes in those perfect examples I collected?" That story starts with a female acorn weevil—large by weevil standards at three-eighths of an inch—that uses her mandibles at the end of a long snout (called a rostrum) to chew a small hole in the developing acorn. She backs up to the hole and deposits an egg that soon hatches into a tiny grub. As the acorn grows, the hole heals over. During the summer, the grub lives inside the acorn, feeding on the growing acorn's flesh, ultimately leaving mostly a hollow shell. When the acorn drops from the tree, this triggers the weevil grub to chew its way out and pupate in the soil, where it spends the winter. Note that the grub has a dark-brown head that is much harder than the rest of the body. That is where you will find the strong mandibles that allowed the larva to eat the acorn's flesh as well as chew its way out. In spring the weevils emerge and mate, and the process starts again.

A question I am often asked is "Where are these 'large weevils'? I've never seen one!" The mystery involves the behavior of adult weevils. They are large enough to be prey items for many creatures, including birds and ground-nesting wasps that provision their nests with paralyzed acorn-weevil adults. To survive, they have developed an innate response to any sort of disturbance—they fold their legs against their body and drop to the ground. To see an adult acorn weevil takes stealth, a diligent search of developing acorns, and a bit of luck. Most encounters are with the grubs. These weevils can be a pest for those organizations that regularly collect acorns for reforestation projects. In some years, nearly all the acorns collected will have acorn weevils living in them, and the infested acorns will not germinate. But for the curious and patient, collect a selection of acorns and keep them in a shallow covered tray of dirt in a cool place over the winter. Next spring you just might be privileged to have that first encounter with an elusive acorn weevil adult.

HIDDEN TIGERS

MICHAEL JEFFORDS

The entomological world is full of analogous organisms, so named because their attitude or demeanor reflects the behavior of larger, showier animals. The tiger beetle—a type of ground beetle—is one insect that lives up to its name. Their common name reflects their reputation for being voracious predators that run down prey, overpower it, and consume it on the spot. While their mammalian counterparts may be more familiar to us, I have no doubt that humans have seen many more tiger beetles than actual tigers. Anyone who has walked a woodland trail will likely have noted very small brilliant-green insects darting just ahead on the path. These are tiger beetles. Sightings are even more frequent among people who visit beaches. While we may think those pristine white-sand stretches are devoid of all creatures, in reality they teem with life, including many species of tiger beetles. On warm, sunny days, scores of beetles will be running up and down the warm sand, searching for prey. We are largely oblivious to theses tiny insects because we often have other nonentomological pursuits in mind while at the beach.

A green tiger beetle is a familiar resident of forest trails in the eastern United States (*top left*). A tiger beetle larva (species undetermined) shows its hunting posture while residing in a tiny glass aquarium (*top right*). A tiger beetle larva demonstrates its prey-capture "leap-out" pose (*bottom left*). Rock-loving tiger beetles mate on a sandstone ledge in southern Illinois (*bottom right*).

I collected tiger beetles as a youth and more recently spent considerable time trying to photograph the adults. This is not an easy task, because the same features that make them great predators—keen eyesight and quickness of foot—make them difficult to approach. Add the fact that you must be lying prone on the forest floor or on the beach, and you see the problem. While I have managed to capture images of several species, it was always of the adults.

The immature stages of tiger beetles, like other beetles and insects with complete metamorphosis, look nothing like the adults. So where are they, and why don't we see any photographs of "baby" tiger beetles? Any entomology textbook will reveal that tiger beetle larvae—always illustrated by drawings—are just as voracious as the adults, but in a unique way. The larvae are modified grubs that live in burrows in the soil. They have a flattened head, with large mandibles, that is used to block the burrow opening. When prey items walk over it or nearby, the larva will quickly reach out from its burrow, snare the prey, and drag it inside. The larva has a modified abdominal segment with hooks that allows it to grasp the side of the burrow so they are difficult to remove, if you should ever find one.

For more than forty years, I photographed insects without ever seeing a tiger beetle larva—until one day in 2014, when my colleague Dr. Joe Spencer called me to his lab. His student, while washing soil samples, had uncovered a tiger beetle larva! What excitement, but how to photograph it in its natural state? After a little thought, we constructed a tiny narrow aquarium only a quarter inch wide using thin glass. We filled it with moist, loose soil and introduced the beetle larva. It just sat there. On a whim, I poked a round hole with a small stick behind the larva, and it promptly backed up into the hole and "assumed the pose," as they say. It was clearly visible through the glass, and we spent an enjoyable morning watching and photographing our tiny, voracious friend.

AN UNLIKELY PAIR

MICHAEL JEFFORDS

For those unfamiliar with the rigors of the food chain, the image shown here may take a bit of explaining. Remember those "dog days" in science class when the scintillating topic of "trophic structure" was the feature of the day, often illustrated with various black-and-white graphics that largely failed to inspire any real interest in the concept? I do, but when confronted with the reality of nature, interest soon picks up. One day I was teaching a field class on aquatic insects to a group of high school teachers, and we were doing the requisite sampling—dipping nets into shallow water, scraping the bottom, dumping the catch into shallow trays, and trying to identify the organisms—maybe not everyone's cup of tea, but an interesting way to showcase the inhabitants of an unseen world. A myriad of creatures swam about the enamel tray—bright-red bloodworms, small snails, copepods and amphipods (relatives of crayfish), and a host of other creeping things. These were mostly tiny, requiring a hand lens to reveal enough detail to begin identification. Undisturbed in their underwater realm, these creatures constituted the highly complex aquatic food web of our textbooks. In the pan, though, they just swam around aimlessly, and we could only speculate on how they possibly interacted.

A large water tiger larva (family Dytiscidae) has captured and is feeding on an equally voracious predator, a dragonfly nymph—the food web in action!

That was true only until my second dip, when I captured the unlikely pair pictured here. They were not small, and there was little doubt as to how they were interacting! Here was a teachable moment, and the educators all gathered around my pan. Whether you prefer the food web, food chain, or food pyramid, scientists have assigned names to each "trophic" level. For example, plants are *primary producers*—they produce food from water, carbon dioxide, and sunshine (called photosynthesis) that all other organisms depend on. The second level contains *primary consumers*, those organisms that feed on green plants and convert plant materials into animal proteins. All rather mundane, but the next levels can get more exciting, as was graphically illustrated by the pair in my collecting tray. The *secondary consumers* are predators (meat eaters) and feed on both primary consumers (the herbivores) and, obviously, on each other. What we had stumbled upon was a "trophic" interaction of epic proportions, at least entomologically speaking. The creature with its jaws piercing the head of the second creature is the larval stage of a predaceous diving beetle (family Dytiscidae) and is known as a water tiger. This highly active, and quite large, larva swims freely and feeds on whatever it can catch and overpower. In this case, the prey—a large dragonfly nymph—was nearly as voracious a predator as the water tiger. Dragonfly nymphs have a unique structure for catching prey, an elongated lower lip that they can rapidly extend to catch prey and bring back to their jaws for feeding. Here, though, the water tiger had surprised an unsuspecting nymph and not only pierced it through its brain, but in doing so also immobilized its prey-catching organ. As it turns out, "trophic structure" is not so boring after all, and it really is a dog-eat-dog world out there.

SUNDEW DRAMA

MICHAEL JEFFORDS

One of the iconic wildflower locations Sue and I had always wanted to visit was the Texas hill country during bluebonnet season. On the way, we stopped at the charismatic Big Thicket Preserve of East Texas, billed by the National Park Service as "a landscape of incredible diversity, from longleaf pine forests to cypress-lined bayous." While we expected to see familiar cypress swamps, one particular habitat caught us completely by surprise, and we spent most of our time on two of its trails—the Pitcher Plant Trail and the Sundew Trail. Both had extensive boardwalks that allowed us easy access to unique bogs.

Two metallic wood-boring beetles interact on the sticky surface of a sundew colony. They are too large and powerful to be trapped by the sticky secretions.

These bizarre communities, populated by large numbers of carnivorous plants, occur along a zone in East Texas where several ecosystems converge. Here, the landscapes of the western dry plains meet the moist lowlands of the Southeast, leading to a mix of plants and animals that would normally not coexist. The low wet areas in the sandy forests support the bogs that were the focus of our attention. We saw massive numbers of yellow trumpet flowers (carnivorous pitcher plants), rising like hungry mouths from the damp, soggy soil, while orchids—orange-fringed orchids, rose pogonia, and grass pink—occasionally broke the yellow monotony. Other plants were less familiar to us—yellow bladderpod, arrowroot, yellow sunnybells, and lance-leaved violet. It was the annual sundew, however, that was most fascinating on a sunny afternoon. The tiny carnivorous plants have leaves decorated with gland-tipped hairs that secrete a sticky substance to trap small insects. When an insect is captured, the leaf folds around the hapless prey and slowly digests it.

Carnivorous plants often grow in nutrient-poor soils, using the nitrogen from insects to supplement their nutritional requirements. On this day, though, a drama that did not exactly follow the script unfolded before us. The air was alive with insects and insect sounds, including giant swallowtail butterflies, bumble bees, and bee-mimicking syrphid flies. But it was a pair of large, metallic wood-boring beetles (family Buprestidae) that stole the show. They outbuzzed all the bees as they circled our heads. What were they doing? The larvae of these creatures inhabit wood and have extremely long life cycles. The adults do not stray far from potential tree hosts. They prefer dying trees or trees under stress, so perhaps they were in this boggy habitat looking for a place to lay eggs on the nearby stunted pines. What happened next was totally unexpected, as the beetle pair landed on the gleaming sundew and began to circle each other, oblivious to the sticky secretions. These large beetles were too powerful to be captured by this diminutive plant carnivore. I could not tell if this was a courting ritual between the sexes or perhaps two males vying for a territory. The pair soon buzzed off into the Texas sun, and the denouement to this little drama occurred elsewhere than on the sundew's glistening stage of death.

FICKLE EVOLUTION

MICHAEL JEFFORDS

Making a new "discovery" was pretty exciting for a freshly minted entomologist, like myself, who was still learning the ropes of his first job as a soybean entomologist. I was often bored with reading about soybeans, and I would usually spend my lunch hour at a nearby railroad track lined with prairie plants, searching for things to photograph. One day in early 1982, I happened upon what I thought was a novel ecological scenario involving two closely related moths in the genus *Schinia*—*S. florida* (the evening primrose moth) and *S. gaurae* (the clouded crimson moth). Both adult moths were cryptically colored to match the pattern and colors on the flowers of their food plant, the morning honeysuckle (*Gaura biennis*). Each adult, however, achieved crypsis in a different pattern.

An eighteenth-century illustration depicts the morning honeysuckle and the clouded crimson moth (*left*; image courtesy of the University of Illinois Library, Rare Book Room). Both the evening primrose moth and its caterpillar (*top*) and the clouded crimson caterpillar and adult (*right*) utilize *Gaura* as a host plant.

Later that same week, I noticed two dissimilar caterpillars feeding on those morning honeysuckle plants. One had pebbly skin and was slow moving and colored to match the part of the plant on which it was feeding—green on foliage or pink on flowers. The other caterpillar was boldly patterned and very active and made no attempt to hide. It took some time, but I was able to identify both caterpillars—the cryptic pink or green one was *S. florida* and the boldly patterned one *S. gaurae*. I found it interesting and fascinating that two species of closely related moths, feeding on the same food plant, would have similar strategies for hiding from predators as adults, but diverge in their evolutionary

patterns as caterpillars. Had I discovered a small quirk of evolution in these two species?

I speculated that the caterpillar of the clouded crimson moth was actually mimicking monarch caterpillars and this was a case of Batesian mimicry. Here, a palatable species mimics a toxic species to gain protection from predators. At the time, this had not been recorded. Excited by these observations, I envisioned a paper or two in a prominent journal. Before leaping to publish, though, I did a thorough literature search to determine if this was a new observation. I found little in the current easily accessible literature, so I engaged our librarian for assistance. One "danger" of looking for information from the University of Illinois Library—one of the largest and most prominent in the United States—is that you will likely find something and in places no one would think of looking. By searching for the word *gaura*, a reference appeared from an unlikely source:

> Smith, James Edward, Sir, 1759–1828. *The Natural History of the Rarer Lepidopterous Insects of Georgia: Including Their Systematic Characters, the Particulars of Their Several Metamorphoses, and the Plants on Which They Feed.* London, 1797.

A quick trip to the rare-book room revealed a leather-bound tome with a beautiful illustration of morning honeysuckle with the clouded crimson adult and larva. Not only was I scooped, but by more than 180 years! However, a study documenting the caterpillar as a mimic of the monarch did not appear in print until 1996. I would have followed up and published that hypothesis had I not been toiling diligently, for the ensuing five years, in the vastness of Illinois soybean fields.

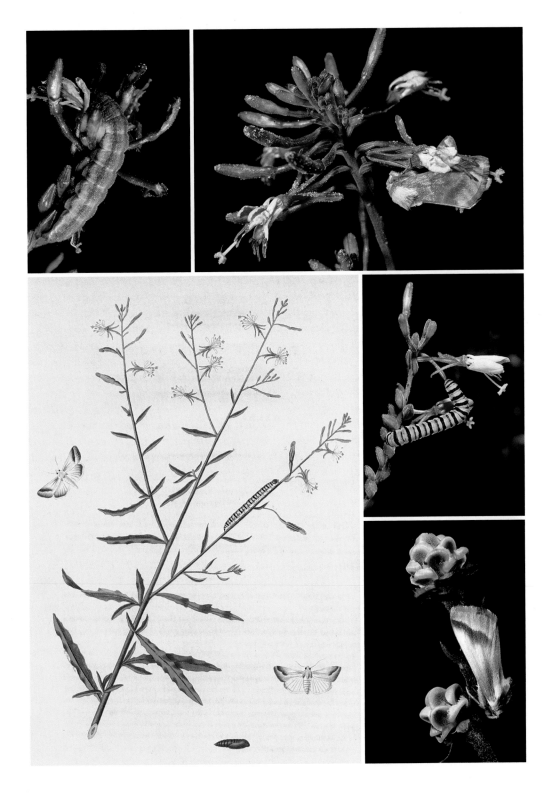

STINK BUG BABIES

MICHAEL JEFFORDS

Being curious means many things to different people—eavesdropping on nearby conversations, sneaking a peak in the oven at tonight's dinner offering, even scanning around the Facebook pages of "friends of friends." To a scientist, curiosity means looking for answers to significant questions that most people have never imagined. The late Wisconsin Democratic senator William Proxmire was famous for his monthly Golden Fleece Award, presented mostly to scientists engaged in research he deemed "a waste of taxpayers' money." Most researchers took a dim view of Senator Proxmire, as it was often basic scientific curiosity that was belittled. Stewart Brand, then editor of *Whole Earth Catalogue*, publicly accused Proxmire of "recklessly attacking legitimate research for the crass purpose of furthering his own political career, with gross indifference as to whether his assertions were true or false as well as the long-term effects on American science and technology policy." The statement was likely true, as Proxmire later apologized for attacking several projects that were canceled.

Newly hatched stink bugs cluster around their egg cases. Hidden on the underside of this leaf in the Sonoran Desert, they soon dispersed to lead a mostly solitary life.

This is all interesting, but what does this have to do with stink bug babies? Directly, very little, but it shows that most science is based on an innate desire to explore and reveal the unknown. While hiking in the Sonoran Desert of Arizona, looking for unusual things to photograph, I randomly turned over a leaf. Underneath was a tiny, hidden world of newly hatched stink bugs clustered around their empty eggshells. I was fascinated by this unlikely nursery, although I did not immediately venture off to submit a grant to the National Science Foundation that might certainly have been subject to *Proxmiring*! What I did do was photograph this small corner of the desert world, and it led me to ask several questions about what was happening. Why were the bugs clustered together? I also noted that they were quite "cute" and wondered if the concept of "cuteness" had any scientific validity. A literature search yielded the detail that some adult stink bugs, after laying a clutch of eggs, actively guard their young against tiny parasitic wasps. Hmmm... Parental care by a lowly insect! While I did not see the mother stink bug, she could have been nearby, as I jostled the leaf when turning it over.

Concerning cuteness, we must turn to Konrad Lorenz, who first introduced the concept as a model in ethology. He stated that *cuteness* is a subjective term that describes traits commonly associated with youth. He called it *Kindchenschema*, or baby schema. Youthful features were thought to make a creature appear "cute" and activate in others the motivation to care for it. Well, it might not have had an effect on the mother stink bug, but I certainly found the diminutive stink bug nymphs charming. I watched them for a few minutes, as they seemed to be feeding on the remaining contents of their eggshells.

I am sure the honorable senator would have thought all this a waste of time. As it happens, I was on a sponsored research trip attending a conference, but I found my trips to the desert much more enlightening and memorable. As for the conference, I don't seem to remember its purpose.

MR. JACKMAN'S BOG

SUSAN POST

Always in search of unusual plants to see and photograph, Michael and I had heard of a bog in northern Indiana that was riddled with rare plants. A bit of research revealed it to be part of the Indiana Dunes National Lakeshore. What we encountered on our visit was most unexpected. Alerted by the barking of his mongrel dogs, the self-appointed keeper of the bog, Mr. Jackman, emerged from his dilapidated house on the isolated gravel road in northern Indiana. The odor emanating from beyond the dark doorway and the humble surroundings bespoke of someone not caring. Not true, as we soon found out the man had a deep attachment for his bog. The informal entrance to Pinhook Bog was through his property—protected by its obscure location and the bevy of barking dogs. In a one-sided conversation in his heavy German accent, he told us of the huckleberries he and his long-deceased wife were renowned for, huckleberries gathered from the bog. They had entertained people from the old country, and guests from as far away as New York came to pick the precious berries. He gleaned them by moonlight, carefully threading his way through the pitcher plants, unaffected by the sticky carpet of sundew. Five years ago, he had discovered that his heavy body was likely to sink in a bog—his boot is still there, somewhere, a fitting testament to the term *boggy soil*. To him,

Pinhook Bog, part of Indiana Dunes National Lakeshore, contains populations of the carnivorous pitcher plant (*left*), pink lady's slipper (*top right*), and orange-fringed orchid (*bottom right*). May and August are the prime times to visit, but only via a guided trip from lakeshore personnel.

though, the bog was full of memories of the bountiful huckleberry harvests shared with wife and friends and a place not to be shared with the casual visitor. After looking us over, he asked, "Would you like to visit?"

After signing in with grizzled Mr. Jackman, we entered a world of quaking surfaces and carnivorous plants, accessed by a precariously narrow boardwalk. Attracting our immediate attention was a spectacular show of orange-fringed orchids—their color reminiscent of the Dreamsicles of our youth. Burnished orange and green pitcher plants rose out of the sphagnum moss, strange, curving periscopes surveying the land for their next meal—a meal that would be caught in the rainwater collected in their pitcher-like leaves. On the velvety green carpet of moss sparkled the crystals of sundew, its sticky secretions catching and reflecting the sun's rays. Dragonfly wings vibrated far above the lethal surfaces. They pursued prey the pitchers had missed and chased down mates to produce eggs to be laid in the dark, quiet water. Remnants of pink lady's slipper orchids that had bloomed in late spring occurred along the wooden path, surrounded by round-stemmed rushes. The narrow boards came to an abrupt end in a thriving sea of sundew, carpeting the treacherous, impassable ground. It was time to retrace our steps.

We visited the same site several years later, but Mr. Jackman, his dogs, and even his house were gone. The bog, with its blooming orchids, was still there, only now protected by a chain-link fence with a formidable-looking padlocked gate. A sign stated that the site was administered by the Indiana Dunes National Lakeshore, with access by "permit only." This was undoubtedly good for the bog, but it was cold comfort for those of us who remembered Mr. Jackman, his sign-in book, and his huckleberries.

AQUATIC CHRISTMAS ORNAMENTS

MICHAEL JEFFORDS

Of all the hidden worlds on this planet, perhaps the least known are the various aquatic systems that cover the majority of the earth's surface. The realm of saltwater accounts for most of this little-known aquatic environment, and we are just beginning to explore and understand the complexities of our oceans. Freshwater systems also play a vital role in human society, most often boiling down to the need for clear freshwater for our consumption. But like the mysterious ocean depths, our rivers, streams, and ponds can be equally as inscrutable. I was reminded of this fact when Sandi, a graduate student from the University of Illinois's Department of Natural Resources and Environmental Sciences, approached me for help with her thesis project—a taxonomic revision of predaceous freshwater mites. I was intrigued, as I knew little about this group of arachnids that are close relatives of spiders. I had encountered them only as dots and specks—most no bigger than a period on this page—swimming around in enamel trays of water samples collected for various projects. They were too small to easily see with the naked eye and had largely gone unnoticed. Sandy needed help photographing these tiny creatures and knew that I was a macro photographer.

Various examples of predaceous aquatic mites from freshwater habitats of the central United States illustrate the tremendous diversity of these tiny arachnids.

We started meeting on a regular basis. After an initial consultation, she began bringing individual mites from her samples to see if I could photograph them.

Here was a challenge: these mites were not only tiny—the largest was the size of a pinhead—but endlessly mobile. They swam incessantly—back and forth and up and down through the water column of the small porcelain container where they were "confined." A new world opened up for me as I trained my 8x macro lens on these minute creatures. I wondered what selective forces drove the evolution of these minuscule predators; regardless, they were spectacular.

Calling them "aquatic Christmas ornaments" is the only descriptor that did them any justice. They ranged in color from brilliant red to the deepest black, with every color of the rainbow in between. Their shapes and conformations were equally as varied and reminded me of everything from black-widow spiders to mutant eggplants and tiny raspberries. There are more than fifteen hundred species of these water demons—most adults are carnivores, scouring their freshwater habitats for any suitably sized prey items. Finding these tiny creatures is not a challenge, as they sometimes occur in incredible numbers and can even be collected in winter, as they forage below the ice.

While most scientific literature states that these are important food sources for small and juvenile fish, it does not account for why they are so brightly colored. Some studies indicate that they may actually taste bad to fish and that their bright colors and decorations—termed aposematic coloration—serve as a warning to leave them alone. The fact that humans notice them only when they are greatly magnified means little. Size is relative. The tiny predacious creatures of aquatic microcosms are subject to the same evolutionary forces that act on all the lions, tigers, and bears that call Earth home.

PSYCHOPHILY

SUSAN POST

Anemophily, entomophily, cantharophily, melittoph-ily, phalaenophily, and *psychophily* flash up on the wall, and I am jolted to attention. When did Michael add this to his presentation? What interesting words, and what do they mean? Michael is giving yet another program, this one on the importance of pollinators, but he has switched things around. These new words are the scientific terms for various types of pol-lination—by wind, insect, beetle, bee, moth, and butterfly, respectively.

A spicebush swallowtail nectars at a rare orange-fringed orchid. Note the pollinia fastened to the front of the head of the butterfly from an earlier visit to another orchid.

As his talk progresses, my mind wanders to a recent field trip to the Iro-quois County Conservation Area. Colleagues from the Illinois Natural History Survey had just returned from fieldwork there and said we had to visit the site soon; the recently discov-ered (in Illinois) orange-fringed orchids were bloom-ing—not one or two, but at least a hundred! With a hand-drawn map and scribbled directions, we headed to a location we had not previously explored. The orange-fringed orchid in Illinois has been called "one of our rarest orchids," and at one time the species was considered extirpated from the state. Yet here we were on this July day, reveling in their Dreamsicle-colored blooms. As if these orchids were not enough, a species of *Spiraea* was also in full bloom, adding accents of pale pink to the diverse shades of green and orange. As I began to photograph the orchids—seeking that perfect specimen to portray—I did not pay attention to the many tiger and spicebush swallowtails that hovered around. As the clouds played hide-and-seek

with the sun, the butterflies, too, came and went. Once the sun reappeared, out came the more numerous spicebush swallowtails. They flew a circuit of the orchids and seemed to stop at the same plants each time. I even began to recognize individual butterflies, due to the different bird-beak bite marks on their wings. I assumed they were nectaring on the orchids and began to pursue several—a photo of a swallowtail butterfly on an orange-fringed orchid would cap an already great day. Back and forth I walked; patience and luck seemed to be on my side, as I managed to get several photos.

On close inspection, I noticed another level of complexity. Sure enough, the spicebush swallowtails were pollinating the orchids; pollinia (pollen bundles that looked like orange knobs) had adhered to the forehead of one of the butterflies! This was a first for me. I was familiar with the eastern prairie white-fringed orchid (pollinated by night-flying sphinx moths), so I had assumed these related orchids (in the same genus and with a similar structure) were also moth pollinated. Obviously, I was wrong. I shared my discovery with Michael, and he also followed the spicebush swallowtails around the orchids.

Near dusk, as we walked back toward our car, we passed through the habitat that ringed the orchid site. It was an "old field" (fallow agricultural land being allowed to revegetate) backed by a savanna on the uplands. Here was the source of our pollinators. The landscape held abundant nectar and food plants—an ideal structure for butterflies. What I had witnessed, indeed, was psychophily by spicebush butterflies. When I finally resurfaced from my "orchid-butterfly reverie," the presentation was nearly completed. I inwardly smiled, proud that I do pay attention to Michael, at least once in a while.

THE THIN ONES

MICHAEL JEFFORDS

Imagine eight pieces of thread attached to a thin wire, add muscles, wings, eyes, mouthparts, and all the internal organs of a "normal" insect, and you have a thread-legged bug, an impossibly bizarre member of the assassin bug family (Reduviidae). Those first encountering this delicate insect often ask, "How can this creature be alive? There's no room for anything!"

A thread-legged bug stakes out a spiderweb for prey (*left*). Three bugs are engaged in a mating sequence. Note the tiny wings visible on the lowest bug (*right*).

Humans are relatively large, as animals go, so we often have a skewed perception of scale. In their hidden world, these voracious predators are perfectly shaped and adapted for their role in the food web. The long antennae are sensitive to the slightest puff of air that alerts the bug to possible incoming prey items. As ambush predators, their front legs resemble very spindly versions of the raptorial forelegs of the praying mantis. Like mantids, their forelegs serve as prey catchers and are not used for walking. The four exceptionally long walking legs provide a stable base for this wispy creature. In flight thread-legged bugs move so slowly that they seem to float along. While they have two pairs of wings, the front pair is modified into thickened structures to protect the delicate hind wings. Often the only things visible are the light areas at leg joints that give the insect an extremely ethereal appearance. These traits sound very intriguing to humans, because thread-legged bugs are too small and fragile to do us any harm. Though they have a very sharp beak, and use it to subdue prey by injecting a potent mix of digestive enzymes and paralyzing agents, they cannot pierce human skin. Any insect of the proper size is fair game, though, and all manner of small flies, caterpillars, and moths fall prey to these delicate "tigers" of the insect world.

In southern Illinois, I have observed a rather bizarre twist on their predatory behavior: some thread-legged bugs use spiderwebs for catching prey. The webs they frequent are not the elegant orb webs of the larger spiders, but the messy webs often found embedded under sandstone bluffs and in thick vegetation. The bugs daintily walk across the surface of the webs, or patrol the edges, carefully placing each of the four walking legs so they do not become entangled in the sticky threads. They walk onto a web and patiently wait for something to happen. Thread-legged bugs are so light and delicate, they do not seem to disturb the web enough to alert the spider of their presence. On two occasions, I witnessed a small moth fly into a bug-inhabited spiderweb, and it was immediately plucked free by the thread-legged bug. It happened so rapidly that the spider was not aware that its pocket had been picked! Trapped in the raptorial front legs, the prey was quickly pierced, and the bug proceeded to suck the digested internal organs, right on the spider's web.

During a late-fall foray, I located what looked like a small tangle of wire on a boardwalk railing in a southern Illinois swamp—three thread-legged bugs were intertwined in a confusing mass of legs and antennae. It turned out to be a mated pair of bugs that were apparently being harassed by a second male.

Anyone who is interested in finding these unique creatures in their forest habitat is most likely to be successful by searching messy spiderwebs or man-made structures like boardwalks and trail signs. Otherwise, this insect may just remain an unseen, thin thread in the fabric of nature.

MOMENTS IN TIME

Only a few things can evoke universal silence from a crowd of humans—a funeral procession, a grand cathedral, or a classroom of children suddenly under the glare of an angry teacher. On a trip to Zion National Park, we experienced another. A trail led to a series of three emerald pools, each encountered sequentially as we stair-stepped our way up a steep-walled sandstone canyon. Emerald Pool #1 was nearly dry, but the sloping undercut wall above it, constantly dripping cool water, was not. It teemed with botanical life—maidenhair ferns followed diagonal cracks, pink and white shooting stars nestled on tiny moist ledges, and, finally, a mysterious orange monkey flower defied gravity as it clung, seemingly to nothing, on the inward sloping wall. Emerald Pool #2 was also nearly dry and quite unremarkable, except as a waypoint to upper Emerald Pool. The third pool was another third of a

An undercut sandstone cliff resides near Emerald Pool #1, Zion National Park, Utah (previous page). A sandstone wall towers above Emerald Pool #3 in Zion National Park (left).

mile up a rocky, sandy difficult trail. Would we go? The question was immediately dismissed, as we, and an eager multitude of others, headed up. Fifteen hot, sweaty minutes later, Emerald Pool #3 came into view, but it's not the pool, per se, that stuns everyone into silence, but the setting. The modest oblong pool is lorded over by vertical sandstone walls of immense proportions, rising at least a thousand skyward feet on three sides. Diagonal cracks running across the lower third of the wall support hanging gardens of ferns, monkey flowers, and yellow columbines. The walls are stained with centuries of dissolved minerals, creating a vertical kaleidoscope of color. While we were there, the only words spoken by the dozens of people were "I feel so small." If the hanging gardens of old Nebuchadnezzar were one of the wonders of the ancient world, then Emerald Pool #3 is certainly a natural wonder of the modern world. For this collection of essays, we have grouped stories about events that impacted us in various ways—most are simply snippets of the natural world that we happened upon, and as biologists they caused us to pause, reflect, and remember.

GRUMPY SPIDERS

MICHAEL JEFFORDS

Few things in nature are as elegant and ephemeral as an orb-weaving spider's web. From a human perspective, these intricately constructed webs are a subject of awe and wonder. It is likely that spiders, however, have a different view, as the creation of a web is a time-consuming, energetically costly endeavor upon which their survival depends. As we all know, these webs are death traps for the many insects that are the food source for the spiders. Orb weavers are some of the largest and most colorful of all spiders and among those most familiar to humans. The quiet corners outside our homes often sport immense webs that seem to magically appear overnight, often with a large garden spider sitting in the middle. Depending on your attitude, this may be a joyous occasion (for an entomologist) or reason to run screaming into the house (for an arachnophobe). Nonetheless, the creation of a spiderweb is a remarkable feat, but what happens to the web over time? Because silk, the material of spiderwebs, is proteinaceous and valuable to the spider, when the web deteriorates, spiders often ball up the old web and ingest the silk to recycle the nutrients. Therefore, old spiderwebs do not go to waste.

An *Argiope* spider occupies her web that has trapped not insect prey, but milkweed seeds! It is not much use to the spider (*left*). A webful of soybean aphids is effectively caught, but, unfortunately, they are too small for the spider to ingest (*right*).

What happens, however, when environmental events occur that are beyond the control of spiders? A simple example is a web that has become coated in morning dew, often a beautiful and surreal sight to us. But water is heavy, and the web stretches and droops. The spider simply waits until it dries and then tightens the web, using many of the structural silk strands that support the web. No problem.

On occasion, however, events occur that are basically impossible for the spider to remedy. As shown here, note a large *Argiope* spider that has built her web in an Illinois prairie in early autumn. Unfortunately, the site she chose was downwind from a mature milkweed plant that was shedding its parachute-like seeds. Many of these seeds had become trapped in the sticky web, never to be released, and certainly of no use to the spider. I watched for a time, and the spider just sat in the middle, seemingly oblivious to the fact that her web was no longer an effective trap. It was now highly visible to flying insects and could easily be avoided. The next morning when I returned, the web was still there, but the spider had moved on.

Another example of orb-weaver frustration occurred near my office in Champaign. All cities in central Illinois are surrounded by thousands of acres of corn and soybeans, and during some years in late summer, the introduced soybean aphid becomes incredibly numerous. As the aphids develop into their winged stage, they take to the air, seeking a host plant on which to lay eggs for the winter. One morning I noticed that a large orb weaver had constructed a pristine web on a shrub outside my window. However, a short time later, the web was highly visible! A close look revealed that hundreds of soybean aphids had become trapped in the web. This could have been a windfall for the spider, were it not for the fact that the aphids were simply too tiny for her to ingest. Imagine, a web full of insects that can't be eaten! All that work was for nothing—a grumpy spider indeed.

FROG RELIEF

MICHAEL JEFFORDS

Frogs, like many other creatures in nature, are in trouble. An increasing number of amphibians are declining in numbers, and some, like the golden toad of Central America, have disappeared altogether. The culprits are many, including man-made chemicals (for example, herbicides) introduced into the environment that ends up in freshwater. Most frogs and toads spend at least part of their lives as aquatic organisms, and their glandular skin absorbs whatever is in the water. This makes them vulnerable to waterborne pollution. Another factor is the worldwide spread of a fungal disease, chytridiomycosis. This disease is linked to population declines and likely extinctions in most parts of the world. If you are an amphibian (frog, toad, or salamander), the long-term outlook for future survival is not very good. As depressing as this is, when I encounter any of these delightful creatures, I quickly relegate their gloomy prospects to the back of my mind.

A bas-relief sculpture of a green frog is created in duckweed. Note the cricket frog perched on the green frog's leg (*top*). A quick-frozen southern leopard frog sculpture was produced by the thin ice (*bottom*).

Over the years, I have observed two most unusual events involving frogs. The first was in a duckweed-covered swamp in southern Illinois that once supported a large, diverse population of frogs. When walking into the swamp on a floating boardwalk, the squeaks and plops of southern leopard and green frogs were a common accompaniment, although their shyness made photography very difficult. One spring day, I encountered a rather large green frog sitting on a log, but it saw me and quickly jumped into the quiet water. Gone. Soon, I noticed that it had quietly floated to the surface a few feet away. However, it was visible only as a frog-shaped bas-relief sculpture in duckweed! Defined as a projecting image of shallow overall depth, bas-relief is most commonly encountered in coins. Here, in the still waters of the swamp, the frog had unwittingly created its own diminutive work of art. A close look at the photo revealed a small cricket frog taking advantage of the shallow perch created by the leopard frog's leg.

During a different season, but at the same location, I witnessed a capricious weather event that may be another factor in frog declines. Late one November, the weather in the morning was quite warm, and this area of southern Illinois had not yet experienced its first frost. Later that same day, an Arctic cold front swept through the swamp, leaving layered sheets of thin ice overlapping on the normally flat swamp waters. I busily photographed this unusual scene, pointing my camera down and recording the intricate patterns. Later, while processing the digital images, I came across another frog also engaged in the creation of a bas-relief sculpture, although the creative process that resulted in this sculpture likely had lasting consequences for the amphibian artist. There, at the edge of the photograph, was a near-perfect relief carving of a southern leopard frog! This time the medium was not duckweed but swamp water that had blown over the creature as it was caught in the cold wind, leaving a quick-frozen frog encased in an icy winter mask.

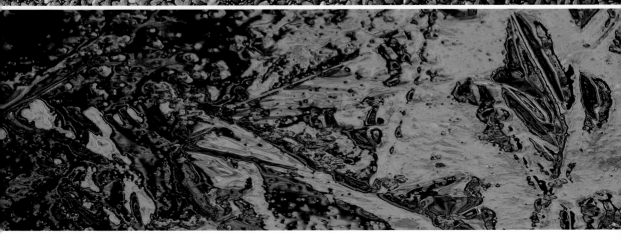

LORIKEET LURE

MICHAEL JEFFORDS

Every ecosystem on Earth has requirements that are integral to its existence. Cloud forests need abundant moisture in the form of humidity and frequent rains; the Arctic tundra requires a layer of permafrost below the surface of the ground that allows a thawed layer of soil above the frost to develop each summer for active plant growth; the prairies of the midwestern United States are maintained by a periodic fire regime. In Australia much of the continent is considered "outback," and, not surprisingly, the limiting commodity here is simply water. Even though this area of Australia receives considerable rain during a year, it is often sporadic, frequently occurring in downpours that cause flooding. The water, however, soon disappears. Consequently, the landscape is semiarid and experiences frequent droughts across much of the area.

Rainbow lorikeets gather at a leaking irrigation tower for an evening drink. Although the prime location was near the top, birds gathered the full length of the pipe for access to the water.

I had this vividly illustrated during a 2011 trip to northeastern Australia. South of Kakadu National Park, on the edge of what is considered the outback (the term actually means remote areas with a sparse human population), we were in a tourist camp that featured small cabins, a cooking pavilion, and several tent campsites. Each of the tent sites came equipped with a water spigot and an irrigation tower, presumably to provide a spot of green grass in the sere landscape. Because it was the off-season, the water towers were turned off, except for one near our pavilion. Even so, it was not exactly spewing water across the campground; rather, it merely dribbled a constant stream down its galvanized expanse. As dinner was being prepared, I noticed a nearby acacia was decorated like a Christmas tree with dozens of rainbow lorikeets. The rainbow lorikeet is perhaps the world's most colorful parrot and a favorite of the pet trade. In the wild, these birds travel in pairs, but occasionally respond to calls and form flocks. Rainbow lorikeets are widespread in Australia, quite common, and inhabit rain forests, woodlands, coastal bush habitats, and even cities and towns. Here at our dry campground, while a river flowed nearby bounded by dense riparian strips, easily accessible water still appeared to be in demand. One by one, the lorikeets from the bare tree noted the leaking irrigation spigot and flew down to imbibe the trickling treasure. Dinner forgotten, our group watched pairs, trios, and sometimes even a dozen of these fabulously colored birds hover, squawk, and drink for a few seconds, only to be chased off by thirsty companions. The show lasted for more than an hour, until just before dark, when we reluctantly put our cameras away for a meal that was now cold and destined to be eaten in the dark.

A GOOD THWONKING!

SUSAN POST

When we travel, I always hope our trips play out like a nature documentary—seeing animal behaviors that are featured in the videos, but witnessed only by those who have spent years "living with the animals." As residents of agricultural central Illinois, we value each day spent in a different landscape, as each is filled with new experiences. Most of the time, we are able to observe at our leisure, often photographing until our hands hurt. This was not the case, however, when we visited the game parks of Africa. Most had evening curfews, just like when we were teenagers. We had to be back within the camp gates before a certain time (usually 6:00 p.m. or at dusk) or were locked out and had to pay a fine to get back in. On the savannas of Africa, this is to prevent wayward explorers from spending a night outside and possibly becoming part of the food chain. So, predictably, when do we have the best lighting and animal observations? It's usually around 5:30 p.m., several kilometers from camp, and on rutted dirt roads—frustrating!

Young male giraffes engage in a "necking" battle for dominance (*right*). As the light fades, both giraffes take a breather (*left*). Who won this encounter? It was still up for grabs when we departed to make our camp-entrance deadline.

One other important rule of the African game parks is that you are not allowed out of your vehicle, for the same reason you must be in before dark. Etosha National Park in Namibia, during their winter (June), is a flat pan, gray-white and limitless. Moisture is found only at water holes. Here the animals gathered, and so did we. The day offered amazing sights—a dusty elephant with a broken tusk plowing through mopane trees like a walking boulder, a small white feather caught in its tail; a distant secretary bird on its nest, silhouetted by the sun; helmeted guinea fowl coming in to drink, resembling a school of angel fish, with their bodies compressed and moving in unison.

We began the long drive back to camp, pausing to photograph forty giraffes clustered at a watering hole—many with young—and splaying out their thin legs in an awkward position to take a tentative drink. We quickly moved on, as there were still kilometers to go. At 5:15 p.m., we came upon more giraffes, but they were not drinking but sparring, or "necking," as mammalogists call it. A pair of giraffes were twisting their dexterous necks and clubbing, swinging, and *thwonking* each other—a dizzying display that sounded like a carpet being beaten. Neck sparring is a preoccupation of young male giraffes, and like so many male activities, it establishes a hierarchy among rivals. It is a test of strength where the giraffe uses its neck and head as a weapon. Our contenders stood side by side, straining against each other, trying to deliver blows to the head. Watching from the sidelines, it appeared like a choreographed dance, necks parallel, then arched or twisted over the other's back, sometimes intertwined. It was beautiful in the fading light, but certainly not serene. Necks and jaws are often broken. By the time males become adults, their territories—indicating hierarchy—are firmly established.

I do not know the outcome of this sparring. We had to leave—the camp gate would soon close. As we sped back to make curfew, spewing gravel and leaving a massive dusty cloud, I shook my head in wonder—did I just witness that? We entered the gate at 5:58 p.m., with two minutes to spare.

SEX ON THE FLY

MICHAEL JEFFORDS

One of the most familiar insects observed during the summer is the massively large green darner dragonfly. While most people do not know its name, the familiar darting flight over playgrounds and parks often sparks momentary interest, perhaps even curiosity. Upon closer examination, this insect proves endlessly fascinating and is peculiarly sexy. Dragonflies belong to the insect order Odonata. Interestingly, they cannot fold their wings and must always hold them perpendicular to the body. They are most often visible while flying or perhaps momentarily resting along the edge of a pond or lakeshore. Based on their extremely large eyes, its easy to deduce that they are predators. They hawk their prey from the sky, as they are incredible aerialists that can outmaneuver just about anything. Green darners can hover, fly up, down, sideways, even backward, and they do it with split-second timing and accuracy. If you're in a darner's field of view and are the appropriate size, chances are you may be on the menu!

A mated pair of green darners oviposits in a quiet pond. Note that the male keeps possession of the female until she has laid all her eggs, fertilized by his sperm (*right*). A mated pair of green darner dragonflies rests on foliage in the "wheel position" (*inset*).

A green darner's eyes are made up of thousands of ommatidia (simple eyes), and their vision is quite acute. Staring into the eyes of a large dragonfly is like looking into the depths of the cosmos—you see neither empathy nor recognition. At darner mating time, their instinctual predatory nature becomes problematic, as an amorous male is certainly the correct size for a hearty meal. All dragonflies are voracious predators that have evolved a unique reproductive mechanism that avoids the male being consumed, not by passion, but by his intended mate. Male dragonflies have two sets of genitals, one located at the end of the abdomen that connects to the internal sperm-producing testes and a second set that consists of an intromittent organ located near the base of his abdomen. On the far right, you can see the blue male has a differently shaped abdomen than the green female. When sexually mature, a male dragonfly will produce a sperm packet and bend his abdomen until his first set of genitals comes in contact with his intromittent organ (that is, the second set of genitals). Here he deposits his sperm packet for later use. Now it's time to go in search of a female. Males are extremely territorial and will patrol, back and forth, along the margins of water, keeping out unwanted, competitive males, all the while watching for females to enter their territory. When a female comes into view, the male darner will fly behind and above her, and when he sees his opportunity, he will fly forward and grasp her behind her head with a special pair of claspers on the tip of his abdomen. Bingo! Once in his grasp, she cannot consume him, and they may fly about, in tandem, until she initiates the mating process. The female will bend her abdomen around in a circle and contact the male's intromittent organ (with the sperm packet) and take it into her reproductive tract. Entomologists call this the "wheel position." Mission accomplished, but the scenario is not over. They may rest for a time, but soon the pair will fly to a quiet body of water, where the male will stay in contact, protecting his sperm investment, while the female inserts the newly fertilized eggs in plant stems near the water's edge. All in all, the process is quite elegant, effective, and downright bizarre!

TOUCHED FROM ABOVE

SUSAN POST

Many years ago, our local public broadcasting station aired a program on exploring Tierra del Fuego at the tip of South America, and one segment featured the striated caracara—also called the Johnny rook—a raptor. These birds are remarkably curious and immediately take possession of anything small left unattended. Anyone in this bird's domain has to remain alert at all times. I kept the Johnny rook in the back of my mind as a bird I had to encounter.

A striated caracara follows Sue, looking for a chance to snatch anything left unattended (*right*). A group fights over leftover tuna fish outside the cabin (*inset*).

Later, during my preparation for a trip to the Falkland Islands, I learned the archipelago is a stronghold for the Johnny rook, with around five hundred breeding pairs. Maybe I would finally get to see one? Our exploration of the Falklands included visits to several outlying islands; the first was Carcass Island. As we settled into our rooms after a short, windy flight, I opened the curtain, and there, staring me in the face, was a Johnny rook. The owner of the island, Rob McGill, explained that most of the Johnny rooks were banded for an ecological study, and he warned us not to lay anything down. Personally, he had lost several watches to the bird's curiosity! After lunch the cook tossed scraps out the back door, and clusters of the birds appeared. Who knew it would be this easy? As we explored the island, several of the birds would follow us, and when I put my camera bag down, one even had the audacity to try to pick it up. Lucky for me, I travel heavy! The Johnny rooks were always watching and waiting for a mislaid hat or lens cap, but I remained vigilant.

Our next island stop found us isolated in a small cabin on Saunders Island. Again we were greeted—and surrounded—by ever-watchful Johnny rooks. I began my island exploration by walking through a sheep pasture pockmarked with Magellanic penguin burrows, when I noticed a shadow above me: a Johnny rook had taken to following me, just waiting for a slipup to confiscate something—unnerving! On this island, the Johnny rooks were apparently more aggressive. At the penguin colonies, the rooks patrol the perimeter, waiting for a chance to snatch an egg or juvenile. The penguin mothers are ever watchful, and as one bird came too close, the chick exposed its rear and squirted Johnny in the face with feces!

During our final day on Saunders, Michael headed a bit farther along the hill for a few more photos of a black-browed albatross colony. I was content to watch rockhopper penguin antics and squabbles. I felt a tap on my head and thought, "Oh, it's Michael, and he is ready to head back to camp." I looked up and was staring into the beady eyes of a Johnny rook, hovering just above my head. I then knew how prey must feel, and I was freaked out. While I love to watch birds, being touched by them is not my thing! I quickly tried evasive shooing and then headed for shelter. Michael was on his own!

That evening, as we were packing up to leave, we tossed a half-eaten can of tuna outside; within seconds three dozen Johnny Rooks were playing "keep away" with the can—grasping it and greedily eating its contents. All through the night, my dreams were punctuated by the bird's feet tap-dancing on the cabin's tin roof, and I imagined their beady eyes watching me. In the morning, a lone bird was sitting on the fence post outside the gate—a sentinel, still hoping that I would drop a shiny bauble.

SANDS OF THE FALKLANDS

MICHAEL JEFFORDS

Some words just don't seem to belong together: *sandstorm* and *penguins* certainly fit the bill. Even so, on a visit to the remote Falkland Islands, I witnessed a unique pairing of the duo. The location was Saunders Island, a privately owned dot of land on the north side of the Falkland archipelago. Saunders Island is approximately fifty square miles in extent, divided into two sections by a narrow strip of land called "the Neck." Here, colonies of gentoo, king, and rockhopper penguins share the beach and rugged upland landscape with colonies of black-browed albatross. Unlike penguins that breed on Antarctic ice, the bird colonies here breed on vegetated slopes (albatross, rockhoppers, and Magellanic penguins) and on the sandy and rocky beaches (gentoo and king penguins). These large colonies are both spectacular and intriguing, as one can spend hours watching the interplay of behaviors that organize the life of large bird colonies—nesting, chick care, interbird squabbles, and, of course, predation. Predators are few, but effective, and include skuas, giant petrels, striated caracaras, and the ubiquitous kelp gulls. All the activity was a paradise for us as photographers and biologists.

Our most visually arresting day, however, was spent not on the Neck but on Cliff Point, on the east side of the island. Aptly named, the steep shoreline dropped more than three hundred feet to meet the South Atlantic on the narrow, rocky coast. Here, rockhopper penguin colonies were interspersed with king

A lone gentoo (*front*) and Magellanic penguin cross the expanse of blowing sand (*top*). Magellanic penguins endure a sandstorm on the way to their colony (*bottom*).

cormorants and nesting aggregations of black-browed albatross. Far below, a narrow beach was backed with both gentoo and Magellanic penguin colonies. On the day we visited, we endured extreme weather that included sleet, rain, sporadic snow, even sunshine, but all modified (and made unpleasant) by sustained fifty-mile-an-hour winds! We spent most of the day photographing the colonies while trying to keep from being blown into the sea. On our trek back to our pickup point, we walked along a cliff above the sandy beach. The views were some of the eeriest and most unusual I have ever experienced. Even though the sand had been soaked by rain earlier that day, the fierce winds were blowing it across the beach in long skeins only a few feet above the surface—about penguin height. We stopped and watched as the gentoos and Magellanics traversed between sea and colony, seemingly oblivious to the sand. We certainly weren't, as a few minutes in that environment would have rendered our cameras useless. Because the blowing sand obscured their feet, the penguins appeared to float across the expanse of beach. Those that did stop to rest were soon drifted over and became mere sandy lumps.

The day made me appreciate how well adapted these creatures are to their environment, including an unexpected use for the nictitating membrane. This structure is a transparent eyelid that can be drawn across the eye to protect it. Penguins have them to protect their eyes in the water, camels possess them for sandstorms, but who knew that penguins would also need them for sand? Given the capricious climate of this remote island, what we observed was most likely not a unique occurrence, but it certainly provided an interesting and surreal photo opportunity I will never forget.

FROLIC IN COLIC

SUSAN POST

In real estate, it's location, location, location. In the natural world, it's location, weather, and timing, and when all three come together—magic happens. During the summers of 2010 and 2011, Michael and I roamed Illinois, searching for and photographing butterflies for a state butterfly field guide. Using field notes, past references, and information from experts, we mapped out a strategy to find as many species as possible. One area that kept surfacing in notes and collections was Iroquois County Conservation Area. We were familiar with the site, as we had visited there during previous springs to see its unusual violets and in late summer for white blazing stars. Twenty-five percent of the blazing-star blooms were often white instead of pink. The area is a mix of unusual habitats—wet, shrub sand prairies, sand savannas, and an old glacial lake bed with a remnant bog. Unfamiliar plant species are found here—sundew, blueberries, and colic root. Our first field guide visit was on June 16, 2010, later than we had hoped—this was the first sunny day after six days of heavy rain. We were looking for hairstreaks, small gray-brown butterflies that use oaks as larval food plants. Most Illinois hairstreaks are only on the wing in June and July. As we drove the sandy roads, I spied great spangled fritillaries dancing around blooming butterfly weed—a good omen. We donned boots and hiked to the shrub prairie, once part of an ancient glacial lake bed, and areas of it are still wet. Among the ferns and false wild indigo, colic root was in full bloom. Related to the lilies, its flowers have

Great spangled fritillaries cluster on colic root in a wet shrub prairie in east-central Illinois.

the appearance of slender white wands. Clustered on these white wands were fritillary butterflies. Not just one or two, but dozens! We found the common great spangled, but also Aphrodite, meadow, silver-bordered, and regal. Illinois has five species of fritillaries that can be regularly seen, and they were all here!

Illinois fritillaries are Halloween-hued butterflies that are divided into two subgroups—greater and lesser. The greater fritillaries (genus *Speyeria*) include great spangled, regal, and Aphrodite. They have one generation per year, and females lay their eggs on or near violets, even though the plants may be done for the season. When the caterpillars hatch, they eat their eggshell, crawl to a protected site in the leaf litter, and enter diapause (a period of suspended development) until spring. The lesser fritillaries (genus *Boloria*) include silver-bordered and meadow. They resemble miniature *Speyeria* fritillaries, also use violets for their caterpillars, yet have multiple generations per year.

We had stumbled upon perfect conditions—the weather was calm, sunny, and warm; the colic root was in peak bloom; and both the greater fritillaries and a new generation of lesser fritillaries were emerging. Lines of five or six bright-orange butterflies would periodically fly past. Colic root was everywhere, and fritillaries clustered on the stems—oftentimes two or three species on a single plant! We witnessed nectaring, courting, and mating. During the afternoon, we counted more than two hundred great spangleds, nearly three dozen Aphrodites, a dozen regal and silver-bordered, three or four meadow fritillaries, and a lone coral hairstreak. After several hours in the wet shrub prairie, attending the "fritillary frolic in the colic," we left with smiles, hundreds of photos, and wet knees.

AN ENCOUNTER WITH EXTINCTION

MICHAEL JEFFORDS

Sue and I once took friends on a day hike in Pine Hills Nature Preserve near Shades State Park in west-central Indiana. Dramatic sandstone hogbacks— high, extremely narrow ridges above nearby Sugar Creek—are the significant feature of this area. While negotiating the trail and helping two small children along, I looked down and found myself face-to-face with a carving of a large bird that resembled a pigeon. The edges of the bird were worn and obviously had not been recently carved. What could this be? Surely, this was not the work of a typical graffiti artist. Sue soon came upon the scene and, being always better informed than I, stated that based on her reading, this was a carving of a passenger pigeon etched into the limestone more than one hundred years ago—an obscure, remote memorial created by an unknown individual for a species that no longer exists. Once numbering in the billions, the passenger pigeon became extinct on September 1, 1914, when Martha, the last living pigeon, silently slipped from her perch at the Cincinnati Zoo.

A simple carving of the extinct passenger pigeon (*Columba livia*) graces a sandstone hogback at Pine Hills Nature Preserve, Indiana.

A second animal encounter, involving an equally extinct species, revolved around the *Tyrannosaurus rex*, named SUE, from the Chicago Field Museum. I saw SUE as she was being prepared for exhibit. She was then a pile of chocolate-colored bones of epic proportions with a monstrous skull that left little doubt as to my fate had we met sixty-five million years ago.

Two species, both extinct, two organisms that I will never encounter except as human-created apparitions, produced a different emotional response in me. SUE sent a delightful chill along my spine as I stared into her blank, uncaring eye sockets. But as I sat on the narrow hogback of Indiana sandstone, staring at the surprisingly humble yet powerful stony depiction of a passenger pigeon, I felt none of the "pride" I usually associate with a distant, ennobled species from the remote past. My initial emotion was sadness for a species that I could have overlapped with, that I might even have photographed. Grief followed, not for this stone carving, but for what it represented—billions of living organisms described by Aldo Leopold as a "traveling blast of life." The passenger pigeon was sent into the finality of extinction not by a profound, unpredictable cosmic event, but by a human society that viewed the landscape and its resources as eminently plunderable and virtually inexhaustible.

I admit to being selfish. Why was I deprived? Why couldn't society have waited? What was gained by the extinction of this species represented by a silent witness incised in the rock before me? My only conclusion was that when viewed as an abstraction, extinction is an inevitable, even acceptable, part of a conscienceless planet. However, when viewed through the reality of an equally conscienceless society, it is certainly difficult to accept and should represent a folly in the minds of all.

THE LEATHER PURSE

SUSAN POST

Did you see that?" asks Michael. "It looks like a leather purse trying to cross the road." We were heading down a busy highway in northern Illinois in early July, scoping out sites for a field trip. We stopped the car, turned around, and headed back. There, slowly crossing the road, was a dinner plate–size turtle. I jumped out and whisked the turtle to the safety of the roadside. In my hands was a large, spiny softshell turtle that was so big, its leathery shell had turned to bone. It looked at me with a leech-encrusted head as I marveled at its olive-green body, pointy nose, and clawed flippers. Instead of letting it resume a slow amble among the flooded cornfields, we decided to drive a few miles to the Des Plaines Conservation Area to release it. Back in the car, I placed the turtle near my feet, where it sat staring at me. At the conservation area, we measured its carapace (fifteen inches), and when released it took off like a flat discus-shaped torpedo. Splash—gone! It was only after the rescue that Michael bothered to tell me that spiny softshell turtles have a nasty bite, just like snapping turtles.

Sue rescues a spiny softshell turtle—a road and a flooded cornfield are not its proper habitats.

Of Illinois's sixty species of reptiles, perhaps the most highly adapted for an aquatic existence are the state's two species of softshell turtles—smooth and spiny. The spiny softshell turtle occurs throughout the state in lakes, sloughs, and mud-bottomed streams, but is most abundant in sand-bottomed rivers. It spends most of its time in well-oxygenated water, either foraging, floating at the surface, or buried in the soft river bottom, with only its head and neck protruding.

The exceptional feature of any turtle is its shell, which is divided into two parts: the carapace (upper shell) and the plastron (bottom part of the shell). Both parts of the shell are made up of bones that are covered with horny, scale-like coverings called scutes. Softshell turtles have an almost circular upper shell, and instead of scutes, both their carapace and their plastron are covered with soft, leathery skin. The carapace bends and the edges droop like a flap over the hole through which the head and neck are withdrawn in time of danger. When removed from the water, softshell turtles resemble gray pancakes; this observation has led to the common names of flapjack or pancake turtles. The spiny softshell turtle is carnivorous, feeding on crayfish, fish, and aquatic insects. When feeding it crawls or swims along the bottom of the water in a random fashion, thrusting its snout under stones or into masses of aquatic vegetation. The species is not without predators. Skunks and raccoons destroy its nest and eat the eggs; fishes, snakes, wading birds, and other turtles consume its young. The biggest problem for the adult, however, is decapitation by fishermen after being hooked on their lines. My July turtle encounter awakened my turtle awareness and added an intimate checkmark to my herpetological life list.

CAPTURING MYSTERY

MICHAEL JEFFORDS

The dirt road ended abruptly in an old cemetery. Sloping off to the south, a field of corn stubble stretched toward the wide floodplain of the Ohio River. The sight was a common one in this area of Illinois, as the annual cycle of flood and silt deposition made this land some of the most fertile in the state. The great river was not visible from this relatively high vantage point; only a line of hazy grayness, uniform in height, spread across the horizon. Trekking across the muddy cornfield was only a prelude of the greater wetness to come. As I moved closer, the gray haze changed into a mass of uniformly spaced boles, each angled slightly and topped with a tangle of branches awaiting a burst of early-spring foliage.

Growing up in southern Illinois, I was accustomed to swampy landscapes such as this, but relatively few times had I had occasion to discover and explore new terrain, untrammeled except for the occasional proximity of a laboring tractor. I was on a soybean-insect sampling mission when I encountered this stretch of river floodplain known as Black Bottoms. At the edge of the swamp, I heard a distant yet distinctive squawking. A colony of pterodactyls or pteranodons immediately came to mind, but only fleetingly, for was I not in the twentieth century? As I entered the buttressed ancient trees—water tupelo and bald cypress—the time period ceased to be important. Only the senses played a role here, and they had become muted, timeless, subject to the mood of this ancient place. The ill-defined squawking soon resolved into a series of focused, raucous calls from messy bundles of twigs, woven into massive nests high above. The great-blue-heron rookery had nearly two hundred nests (each with a chick), all evenly spaced and ancient in their own timescale. At the core of the rookery, the ground became a speckled mass of white, gray, and tan—remains of the fishy meals enjoyed above. To demonstrate their displeasure at my proximity, a slight whistling sound from above caused me to look up, just in time to see a partially digested hand-size bluegill come hurtling down and plop wetly at my feet. As I retreated, the eerie call of a barred owl, somewhere in the distance, was answered by one not so far away; both seemed to serenade, but more likely applaud, my departure. I will always remember that visit in the early 1980s and was glad I had carried my camera along.

A few years later, I returned to that same site during high water and canoed through the trees. The herons had not yet returned, but the nests were still waiting, high overhead. A final visit in the 1990s proved to be much less satisfying. The view from the cemetery parking lot now showed only a uniform expanse of farmland, stretching to the brown ribbon of the Ohio River. The land of the herons and barred owls was no more.

Later, I happened upon a couple of the images from that first visit. I was struck that perhaps the only records that this place had ever existed resided within the pockets of my polyethylene slide files and in the now meaningless name "Black Bottoms" from a topo map. I've often sought the reasons I have such a passion for photography, and I think this episode may best epitomize them. Photographing landscapes that are ancient and undisturbed is like recording history, or, in the case of this special place, capturing the mystery of a place too perfect for its own survival.

A bald cypress—water tupelo swamp, known as Black Bottoms, resided on the floodplain of the nearby Ohio River in southern Illinois (*right*). A massive great-blue-heron rookery graced the treetops for many years (*inset*). Both are now gone.

THE BLUE BEECH

MICHAEL JEFFORDS

Trees played an integral part in my early childhood development, but almost entirely a physical one. In those days, I saw trees not as biological entities, but as rough, challenging companions. My yard in southern Illinois had several eminently climbable silver poplars, including one grand example with a large, wide fork several feet off the ground. This perch afforded myself and neighbor kids a secure platform to leisurely while away many a summer afternoon, planning all sorts of activities that engage the minds of young boys.

A magnificent specimen of a blue beech (*Carpinus caroliniana*) once graced the floor of a box canyon in Ferne Clyffe State Park, Illinois.

Besides trees, I had a friend, Jack, who was a few years older and had a 1947 Chevy that we used to go exploring. One spring Sunday in the mid-1960s, we struck out on a particularly adventuresome trip, some thirty miles from home, to a never-before-visited state park called Ferne Clyffe, near Goreville. My recollection of the day was that it took an inordinately long time to get there, required nearly a dollar's worth of gas, and exposed us to wonderful rock formations, deep canyons, and cool valleys. By this time, I had already entered the insect phase of my childhood and had little interest in plants. Trees, however, still held an irresistible pull that said "Climb me." On a short hike in a deep canyon, we encountered one particular trailside tree near a cool, rocky stream. The shore swarmed with black-winged damselflies. The tree caught my attention because its bark was a deep blue, and the trunk and branches were seemingly corded with muscles! It was one that I had never see before, and I had no idea what species it was. Although not large by my silver-poplar standards, it had a wide-spreading canopy and an imposing presence. What a body that tree had—one certainly the envy of a tall, spindly fourteen-year-old! After several useless trips through my encyclopedias, I later found a *Golden Guide to Trees* that listed my tree as a blue beech or ironwood (*Carpinus caroliniana*). As my friend and I continued to explore for the next few years, I often encountered other examples of blue beech, particularly along streams, but none so impressive as the one I first encountered.

Not too many years ago, Sue and I were hiking the park trails of Ferne Clyffe for a book she was writing, when what should materialize along a misty canyon trail but that same blue beech! I had nearly forgotten about that old friend. Since our last encounter, we (the tree and I) had both grown considerably, although it still had a better body than I did. But now that blue beech, still relatively small compared to an oak, was otherwise absolutely magnificent! Could this be a state champion? This time I photographed it, so its image would not fade from my memory again. Surprisingly, Illinois's state parks were somewhat flush with funds in 2010, and many upgraded their trails. Imagine my chagrin when I revisited Ferne Clyffe one day and found massive earthmoving equipment around my tree. Its lower branches had been chopped off to make way for the new trail that now imprisoned this once magnificent specimen. The tree survived for a few more years, but died in 2014. It now exists only as a decaying log, covered in shelf fungi, alongside the new and "improved" trail.

PHYSICAL PHENOMENA

At times during our careers, we have collaborated with geological colleagues, especially on outreach activities. This can be challenging, because biologists and geologists often have very different perspectives on the natural world, and sometimes we are sure it sounds like a game of geology versus biology. This was not a comfortable scenario for some audiences that failed to recognize that we were merely jousting, mostly in jest. It was frustrating for all involved when people missed the connections between our disciplines. Thus, we soon developed a "catchphrase" that seemed to satisfy both our disciplines and our audiences: *Geology dictates biodiversity and biodiversity modifies geology*. The phrase captures the synergy between the physical and the biological. The interface between these inextricably linked yet separate worlds is perhaps one of the most interesting facets of science. We witnessed an example of the connection between these two forces in the American West, where the cloak of vegetation fails to mask the unique interactions between the physical and the biological worlds.

A young raccoon safely bedded down in a hollow tree was rudely awakened when an ice storm brought down its home in the middle of a swamp (*previous page*)! An ancient bristlecone pine clings to life in an extremely hostile landscape (*left*).

Absolute quiet reigns here (Cedar Breaks National Monument, Utah) with not even the buzzing of bees. Boulders and the spectacular rock formations make no sound. It's enough for them merely to exist. While boulders cannot speak, they can still tell us voluminous tales about past events. Granite boulders in Illinois reveal the history of glaciers. Basalt boulders littering Capitol Reef National Park provide evidence of nearby volcanic activity and the erosive power of glacial meltwaters. A vast landscape of jagged, black lava boulders, seemingly cooled only yesterday, makes us aware of the ultimate power that slumbers not far below us. Found in many locations on these pure, barren rocks of the American West grows a remarkable tree—the bristlecone pine. At sixteen hundred years old, the example before us is just approaching early to middle age. From the perspective of the boulders that cradle the base of the tree, it's very young, even though it has been around to witness a significant part of human history. It's the tree that chooses to live where nothing else can. In places, though, we find that even the ancient bristlecones are struggling. Long-dead, twisted, snarling skeletons of tree branches have been nearly freeze-dried by the cold, incessant winds. But the trees have drawn the line, and portions of them still cling to life. The dead limbs, though, have merged with the boulders, and will erode no further; for the wind, it's like whittling on iron. Now this is aging with dignity.

In this series of essays, we are bemused by the quirks of erosion, baffled by the physics of light, made aware of the underlying geology showcased as seasons change, and awed by the span of time necessary for the creation of a simple fossil.

HIDDEN RAINBOWS

MICHAEL JEFFORDS

To see a rainbow, we search the sunny sky during a quick rain shower. If we're lucky, and in the right place when coronal mass ejections occur (a release of gas and electromagnetic energy) from the sun, the aurora borealis (northern lights) may also put on a sky show for our viewing pleasure. But who visits a southern swamp to see rainbows? No one that I know of, yet late one winter day I was privileged (and lucky) to view a remarkable optical phenomenon. One of my favorite places to visit and photograph is Heron Pond State Natural Area in southern Illinois. An early-morning March hike had yielded the typical swamp images of buttressed, straight-trunked bald-cypress trees rising from the quiet waters of the swamp. A green tinge of early duckweed covered the still waters—beautiful, but unremarkable. Later in the day, a freak cold front swept across southern Illinois, dropping the temperature from a balmy sixty degrees to just below freezing. Around dusk I revisited Heron Pond on the off chance that the cold spell might have stimulated wildlife action. I made the two-mile trek for a second time that day. The swamp was different, but not dramatically so, as the surface had acquired a thin sheet of ice, peculiarly layered by the action of the strong winds that had accompanied

Early morning in the swamp (top) was beautiful, but nothing special. After a late-afternoon cold front had swept through, a polarizing filter revealed rainbow patterns in the ice (bottom).

the cold front. On a whim, I placed a polarizing filter over my lens (these filters serve to darken the sky, but will also suppress glare found on the surface of water or, in this case, ice). The change was both dramatic and incredible. In fact, I was stunned to silence by the change in the surface ice. I was surrounded by rainbows, thousands of them, interacting in a colorful kaleidoscope across the frozen face of the swamp. My only thought was that a rainbow had shattered and fallen into the swamp.

An initial analysis was that the erratic freezing patterns of the surface ice had created a complex set of prisms. The uniqueness of this once-in-a-lifetime optical phenomenon was worthy of Isaac Newton himself, who first described the spectrum of sunlight formed by a prism. Later, while delving more deeply into the causes of rainbows, I became embroiled and bewildered by the physics of refracted light—the Huygens principle regarding light entering a fragmented medium at various angles, refractive indices, Snell's law (the ratio of refractive indices), and, my personal favorite, the mathematics of deviation angles and dispersion!

I happily returned to the aesthetic side of my experience and just marveled at the power of light when recorded by my camera. Simply by adding a common filter to my wide-angle lens, a new, invisible world had appeared. Perhaps that is why I am passionate about photography. In many ways, the camera sees the world as we do, but sometimes it offers views into a realm that cannot even be imagined. Complex optics aside, the experience was pure magic.

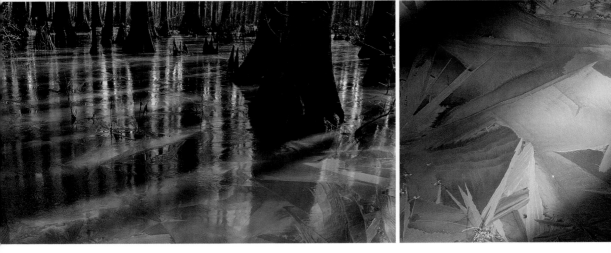

SEASONAL LINES

MICHAEL JEFFORDS

The legendary Maginot line was designed to protect France from an invasion from the East, a fate similar to what it had suffered in World War I at the hands of neighboring Germany. It failed, however, when the Germans invaded through Belgium in 1940. Nature, too, constructs lines, some relatively sharp and obvious (for example, the tree line in high mountains), but others that are more subtle and seldom noticed by anyone. In many years of roaming about Illinois and the Midwest, these natural lines have sometimes come to light, most often dramatized by the changing of the seasons. Lines in the landscape are especially visible in late summer and early fall, when the startling changes in leaf color highlight their delineation. They are certainly present in spring, but the subtlety of various shades of green often masks their presence. Why are these lines important, and what do they mean to biologists, especially those devoted to habitat preservation and restoration?

Both images illustrate the importance of physical features (soil type, moisture regime, and elevation) on the type of forest that can exist. These subtle lines become quite obvious during the fall season.

Most landscapes in the United States, and the world, have been altered by humans. We have harvested forests, converted prairies into agricultural lands, altered stream channels, and attempted many other activities, all to make way for our species and its needs. However, beginning in the 1960s, a movement began across the United States to find and preserve remnant habitats representative of what was here before European settlement. Also included in preservation efforts were areas that still retained some natural features but were in need of restoration activities. The two landscapes illustrated here are examples of subtleties that must be considered by restoration ecologists; they also make for dramatic viewing by those who enjoy quiet walks in nature. Both are forested landscapes along streams, which helps to define the delineation in the habitats. The top image is at the confluence of Dutchman Creek (*right*) and the Cache River (*left*); it's not the rivers we are interested in, but rather the forests. Note the difference in seasons between the forest on the left and the forest on the right. It looks as though summer has stalled in the green forest, while the seasons have progressed on the right. Actually, these are two separate forest types, occupying varying terrains, having dissimilar soils, different tree species, and growing on differing elevations. The same circumstances define the bottom image, found at Beall Woods State Park and Natural Area, except the apparent seasons are reversed in this image. In both instances, the fall sides of the image are mixed mesophytic (moist) hardwood forests composed largely of maples and scattered oaks. The summer sides are both floodplain forests, much wetter, with trees adapted to periodic inundation by the streams. They are made up of elms, ashes, sycamores, and silver maple. The diverse sets of widely varying conditions modify the coming of fall (triggered mostly, but not entirely, by photoperiod) and create very distinct lines—lines that are certainly relevant to ecologists who seek to restore habitats in this ever-changing world.

TOMORROW'S FOSSIL

MICHAEL JEFFORDS

Like many children, I collected fossils. They were relatively easy to find among the limestone riprap along the Ohio River and were a great addition to my miniature five-cent-per-visit back-porch museum. Later, in the 1990s, while teaching a wetlands workshop on land bordering the Illinois River in central Illinois, I happened upon a dobsonfly (order Megaloptera) wing lightly embedded in the silt of a recent flood. I took a photo and promptly forgot about the incident. Many years later, I was discussing fossils and fossil insects with a colleague, Dr. Sam Heads, who is an insect paleontologist. It wasn't long into the conversation before I remembered taking this image and innocently inquired whether this could, indeed, be a fossil in the making—a protofossil? I was gratified that the answer was a resounding yes, but there the simplicity ended.

A dobsonfly wing (order Megaloptera) embedded in the silt left by the receding floodwaters of the Illinois River is likely a fossil in the making.

Below I will try to summarize Sam's explanation of this interesting image. It could be the beginning of an organic compression fossil, provided this was a part of the floodplain where significant deposition occurred (it was) and assuming that the specimen would be buried quickly (the river did flood later that summer). So far, it was looking good for future generations of fossil collectors. But to continue, insect remains are commonly encountered during archaeological excavations, and most experts consider them not to be true fossils, as they are not old enough. Pleistocene material (2.6 million to 12,000 years ago), however, is generally considered to be fossil. "But," I persisted, "how would this wing actually become a fossil?" Consequently, the type of mineralization, the chemistry of the sediment, whether it is oxic (has oxygen) or anoxic (no oxygen), and how much and what kind of pore water (the water occupying the spaces between sediment particles) is moving through the sediment all play critical roles. As expected, it was getting more complicated. Another factor to consider was the degree of decay of the specimen. In this case, the dobsonfly wing is composed mostly of chitin (a derivative of glucose called N-acetylglucosamine) that is quite stable and resistant to decay. Given anoxic conditions and the accumulated factors discussed above, a certain type of bacteria can transform sulfur and iron in the sediments into compounds like pyrite, marcasite, and mackinawite. These will cover or replace (or both) the tissues in the wing, thereby fossilizing it! I asked how long this process would take, that is, lithification (turning to stone) of the dragonfly wing. Surprisingly, it would not need to be buried under a sea for this to happen, but burial in some type of sediments would certainly be required. The depth of burial is often important, but is not the only factor.

To conclude, yes, my dobsonfly wing could perhaps one day become a fossil, but one that is only likely to be found by individuals, perhaps an interested child, a thousand or more generations into the future. Given the number of factors that must come together to create a fossil, I now more fully appreciate fossils and the processes they must undergo to form. Perhaps I should have charged more than a nickel to visit my early-childhood museum.

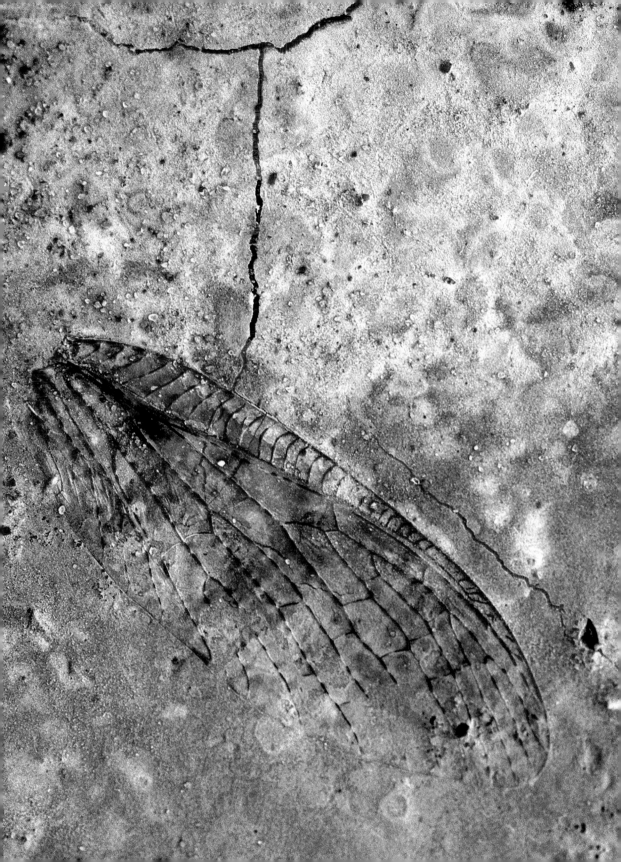

GOBLINS

MICHAEL JEFFORDS WITH SUSAN POST

My first thought on visiting Goblin State Park in remote southern Utah was that it is "a place that illustrates the point that geology has much more of a sense of humor than geologists!" Otherwise, how could this place even exist? In my journal, I noted the area resembled "a collection of some ancient, giant ogre—massive numbers of malformed, misshapen potatoes, each proudly displayed on a small pedestal of mud." All humor aside, the technical term for the bizarre formations in this small, isolated landscape is *hoodoos*. Hoodoos form when a harder layer of sandstone lies atop one that is much softer. The underlying rock erodes faster and gives rise to the pedestals that support the mushroom-shaped rocks that litter the valley floor. This desert landscape is part of the San Rafael Swell, an area of unique geology with features that are reminiscent of those found on the planet Mars. In fact, the Mars Society (an organization dedicated to promoting the human exploration of the Red Planet and works diligently to educate the public on the benefits of the exploration of Mars) has set up a field station near here with a simulated Mars Surface Exploration Habitat. We did not visit their station, as just being here was otherworldly enough.

The main culprit in the formation of the goblins

The landscape of Goblin State Park, Utah (*top*), is both wild and bizarre. An individual Entrada Sandstone hoodoo rests on its pedestal of mudstone (*bottom*).

is Entrada Sandstone, of the Jurassic period, around 170 million years old. Entrada Sandstone consists of debris eroded from the surrounding highlands and deposited on various layers of sandstone, shale, and siltstone. It existed on the edge of a primordial sea and was exposed to the vagaries of ancient tidal streams. The Entrada experienced differential erosion that exploited the fractures in the rock, resulting in the round, potato-shaped formations we see today.

While writing this, it occurred to me to check Sue's journal on this trip, as she is often much more poetic than I am. Here are her observations on this landscape.

> I have encountered a strange, walled city that seems to be inhabited by silent, giant potatoes on pedestals. They have multiple slits for eyes, are forever silent, and just seem to observe. ... Michael has said throughout this trip across Utah that this is what Mars must look like. Today, I think we have landed on the Red Planet and now wander amidst its inhabitants, hopefully undetected. All shapes, sizes, and forms, beyond my imagination to conjure—pagodas, ducks, mushrooms, turtles, and, yes, human faces, all seemingly amused by the capricious nature of the geological gods.

We have twice returned to Goblin State Park since our first encounter and still are in awe of the forces of nature—differential erosion of ancient deposits—that conspired to create this unearthly landscape from simple sandstone.

SUPERIOR ICE

MICHAEL JEFFORDS

Who goes to northern Minnesota (north of Duluth) during a freak cold snap where the temperatures hovered around negative 54°F? No one in their right mind, but we did. A unique ornithological phenomenon—an invasion of great gray owls—proved irresistible. While there we witnessed a fascinating but purely physical event, ice circles along the shore of Lake Superior. The first surprise was that in these frigid temperatures, the lake was still basically unfrozen, with the exception of the relatively quiet waters along the shoreline. This area was protected from lake waves by natural rocky break-waters that created mostly ice-covered bays. Second, we found literally hundreds of perfectly circular ice floes, floating like giant aspirin tablets on the gently rocking waters at the heads of the bays. As a scientist, my first thought was wondering how these could form, but no logical explanation immediately came to mind. The edges of most were slightly raised, so I surmised that the circular form must have something to do with bumping together, but why circles, and, more important, why weren't these quiet water bays frozen solid in the biting cold?

Being a member of a large scientific institute has its advantages, so I contacted colleague James Angel of the Illinois State Water Survey, Prairie Research Institute, University of Illinois, with the above questions. His interpretation was brief, but apparently spot-on, as he said, "I believe this is called pancake ice. My understanding is that in the first stages of ice

Pancake-ice formations float on a quiet inlet on the shore of Lake Superior, north of Duluth, Minnesota.

formation, the ice can be slushy and easily disturbed by any wave action. As it breaks up, it bumps into other pieces, thereby developing rounded and slightly raised edges. It's apparently very common in the polar regions." That certainly made sense, but Duluth, Minnesota, is not really considered polar (albeit the temperature sure made it feel so), although now I had a handle to investigate further. Coincidentally, someone has actually measured "pancake ice" and found the circles range from thirty centimeters to three meters in diameter and can be up to ten centimeters in thickness. Sounds like a long, cold endeavor, but I am glad someone has done it.

This first stage of ice formation mentioned by Angel is called "grease ice," a thin, soupy layer of slush that can resemble an oil slick, and it seems, if conditions are right, that pancake ice is the result. I was also somewhat amused to later find out that local television stations in the northern United States actually forecast the potential formation of pancake ice on the Great Lakes. One such forecast was based on a tweet from the United State Fish and Wildlife's Midwest Region. Who knew? This is possibly the second oddest bit of weather predicting I have run across, surpassed only by the "sheep chill index" regularly broadcast throughout the Falkland Islands. It turns out that sheep farmers need to know this to time their shearing activities to ensure maximum ovine comfort.

Being adventurous, I tried picking up a circle of pancake ice, certainly contraindicated in negative 50°F temperatures. It was very slushy, and virtually impossible. For the record, I did poke one with my tripod leg, but left only a small dent—not nearly as exciting as the trio of great gray owls watching me from the nearby forest.

LIGHTS, ACTION, CAMERA!

SUSAN POST

As entomologists, one of our favorite activities is looking for insects at night. While it sounds odd, streetlights, gas-station lights, or even a portable UV black light provides hours of entertainment and entomological enlightenment. Insects are attracted to light—they are positively phototactic—but science has yet to determine exactly why. Artificial light sources may disrupt insect navigation, since the moon and stars act as distant, but unreachable, natural navigational landmarks for insects. Artificial lights are close, so insects end up endlessly circling them. Some think lights may indicate a false clear path for flight. Others postulate that insects are fooled into thinking the UV wavelength of light is a flower and thus a food source. No one knows the correct answer.

An immense giant water bug dwarfs a large giant water bug in the Pantanal of Brazil. While both were treasures, one was almost the size of a flip-flop!

We travel to southern Illinois quite a lot during the spring and summer. Most people associate Garden of the Gods, with its beloved Camel Rock, or the entirety of the Shawnee National Forest, as a favorite destination. For us, it is a local gas station, just off Interstate 57, rife with bright lights. From April to August, we find a cornucopia of great insects each night and the next morning. Locals on a perpetual cigarette break sit under a sweet-gum tree and shake their heads as they watch us walk slowing around the building, picking off insect treasures from the facade. Many times, a stop will yield only remnant wings, a plethora of mayflies, or scavenging beetles, but more often than not we find a true gem—luna moths, mating tulip-tree silk moths, giant stag beetles, and, perhaps the ultimate prize, a unicorn beetle.

We check the local lights even when we journey to other countries. In the upper Pantanal of Brazil, we were the only people outside our lodge at night. Jaguars prowled the area, but of that we were blissfully unaware. Mole crickets, however, were flying around us like the monkeys from *The Wizard of Oz*. I found a giant water bug, two inches long, and Michael happily photographed it. I continued to look and soon called out, "Forget that water bug. I have found something better!" The prize was an immense water bug more than six inches long. Now that was an insect. Later in that trip, imagine our glee when we arrived at a lodge in the jungles of Brazil that had a permanent black-light setup. During our stay, we felt like kids at Christmas, constantly checking and rechecking the light to see what was new. Throughout the night, different insect species arrived at different times.

After seeing the possibilities of tropical black lighting, in 2013 we scheduled a trip to Ecuador and convinced the owner of the lodge where we were staying—Tandayapa in the Ecuadoran cloud forest—to set up a black light for us. The first morning we were up at four to check the sheet. It was covered with hundreds of yellow and black moths, many subtly different. Could some of these be new species, unknown to science? We also learned an important lesson about black lighting in the tropics: we needed to rise early, as the local bird fauna soon learned to treat it like a bird-feeding buffet.

SELDOM WITNESSED

It is not often that we stumble upon a scene, insert ourselves into the midst of the event, and are able to interpret what has happened and what will occur in the future. It is rarer still to come upon a momentous event at its inception and then to be around to observe the final outcome. Both scenarios occurred at Emiquon Preserve, a Nature Conservancy restoration project, near Havana, Illinois, beginning on September 12, 2004, at 3:15 p.m.

It is a blue-sky late-summer day. A hot, dry wind is blowing. The temperature at the bank in Havana reads 92°F. We cross the Illinois River, turn north, and find endless fields of mature corn, protected from the Illinois River by the high berm of a levee. Gray clouds of dust in the distance mark the location of our destinations—a cornfield being harvested.

On one side of the levee, we hear the buzz of katydids; on the other is the man-made drone of the final harvest. Two dusty John Deere eight-row combines, four equally grimy John Deere and Caterpillar tractors with wagons, and many semis from the local grain elevator are queued up, waiting their turn to load and roll through the field of bone-dry corn stubble. Some of the tractors bear more than a passing resemblance to the old-time dredges that made farming this field possible.

The combines sweep across the field in tandem, like the fabled prairie fires, consuming all in their path. They leave not charred ground, but soil littered with stripped, toppled cornstalks; empty, papery corn husks; and bits and pieces of brickred corncobs. A whirlwind of chaff spews out of the back of the combine like a mini tornado. The debris twirls in the air before coating the remnant cornstalks. Two rabbits flee as the roaring monsters come closer.

The tractors and wagons parallel the combines in a ballet of pick, shell, and pour. The orange corn moves like a flame, twirling and swirling as it goes from combine to wagon. Once each wagon is full, the tractor heads for the waiting empty semis. The corn sounds like rain on a tin roof, pattering and pinging, as it pours into the large truck.

Finally, there is only silence. Long, dried pieces of corn leaves are caught in a whirlwind—revolving and circling upward. The field is harvested, and the crew has moved on while there is still daylight. Time is wasting, as the elevator is open until only five o'clock on Sunday.

A great blue heron flies low over the fresh stubble, while a tan mayfly with extra-long tail filaments alights on a broken cornstalk. The flat, unending cornfield is gone, leaving a landscape of infinite possibilities. This final harvest means that the restoration of Emiquon has begun.

A scant five years later, the same landscape, still flat, is coated not with corn stubble, but with a multitude of birds that often blot out the sky with their restless comings and goings. What we once knew—from our history books—as a backwater lake, explored by Jolliet and Marquette in the seventeenth century and experienced as a central Illinois cornfield for many decades, is now, once again, the wildlife paradise of those early explorers and their American Indian predecessors. Emiquon has come full circle.

The events we showcase in this chapter, while perhaps not as dramatic as the restoration of Emiquon by the Nature Conservancy, are still incredible manifestations of the natural world, but on a widely divergent scale. From a tiny skipper, perhaps flying for the last time, to a multitude of gulls playing on an early winter's day, they are all special.

CORMORANT ROOKERY

MICHAEL JEFFORDS

Aerial photography, the old-fashioned kind where you open the window of a small plane and lean out to take pictures while dealing with hundred-mile-an-hour winds, sounds exciting. It really is, depending on what's on the ground. I have done this many times, but always over my home state, Illinois. Most of Illinois is an agricultural checkerboard, stretching to the horizon in all directions—uniformly unexciting. On occasion, however, the small plane passed over features that both interested and stumped me, at least for a time. Such an event occurred early one spring while traversing the upper end of Carlyle Lake (the part of the lake farthest from the dam), a massive impoundment of twenty-six thousand acres (the largest in the state). The upper end is a mosaic of shallow wetlands and open water, interspersed with many small islands. From three thousand feet, each island appeared as a dot of land, mostly wooded and showing the first suffusion of spring green. One very small island, populated by trees engulfed in the quiet backwater flood, was different. Most were a uniform, stark white—ghostly skeletons that contrasted sharply with the gray-green water. My first thought was that they had been killed by a previous cycle of high water, but the surrounding green trees suggested a different scenario.

An aerial view of a cormorant nesting island in Carlyle Lake (*top*) provides few clues to what is happening at ground level. Ground-level view of a double-crested cormorant roost tree (*bottom*) reveals the bird's identity. Note the stunning green eyes of a double-crested cormorant (*inset*).

Although he was somewhat reluctant, I coaxed the pilot to fly lower and make another pass. Aircraft flying low over large bodies of water are vulnerable to the numerous flocks of Canada geese that call any large impoundment home. But his curiosity must have also been piqued, because down we went, circling over the dead trees. Suddenly, tiny black dots became visible in the branches, hundreds, even thousands, of them. Something clicked (besides my shutter), as I had seen this phenomenon before, in Florida and in northern Illinois. We were viewing, from an unusual perspective for humans, a massive double-crested cormorant rookery!

This widely distributed waterbird is strictly a fish eater, much to the chagrin of both anglers and fish-farm managers, and hunts by swimming and diving. Their numbers were once threatened by DDT, but the population has increased dramatically in the past few decades. Double-crested cormorants build stick nests in trees, often on islands where they are not disturbed, and the massive buildup of acidic guano often results in the death of the trees. This explained the white skeleton-like appearance of the trees on my island. The pilot dared not fly any closer (I would have liked to descend farther to have gotten an estimate of the number of nests), but having a disturbed cormorant fly through a window or into the plane's engine was an event I did not want to experience.

Since then, I have had the opportunity to see and photograph many cormorant colonies around the world (from the ground), but none has had the mystery and charm of that isolated human-created island populated by scores of green-eyed double-crested cormorants.

ALWAYS AT TRAIL'S END

SUSAN POST

One of my favorite books is *North with the Spring*, by Edwin Way Teale. Written in 1951, it sparked my curiosity to see the landscapes he did, particularly Great Smoky Mountains National Park. I wanted to experience the "home of the richest flora and the most luxuriant deciduous forests on the North American continent." Biological diversity is the hallmark of the park. In its eight hundred square miles in the southern Appalachian Mountains, GSMNP has more than twenty-eight hundred species of plants.

The painted trillium is a long-lived wildflower, requiring highly acidic soils. It is widely distributed—but often difficult to find—in the eastern United States (*right*). A Jordan's salamander conceals itself during the day under rocks and rotten logs (*inset*).

While Teale's book did not provide any exact locations for species, I had a copy of *Hiking Trails of the Smokies*, published by the park's Natural History Association. The book listed trails and provided a description of plants and their locations. During our first visit, we met an enthusiastic park ranger, and she shared plant locations with us. Our goal for this trip was to see as many trilliums as possible. Ten species are found here, and after locating some of the more common species—great white, red, and yellow trilliums—we were on the hunt for the painted trillium. Like the other trilliums, we expected to find expanses. Had the book *Trilliums* by Frederick W. Case been published, we would have known that painted trilliums are usually found as scattered individuals, not large clumps, and that they bloom later than other species. But sometimes naïveté is a blessing. We headed to the Porter Creek Trail, known for wildflowers lining the path,

and from our guidebook noted that "painted trillium blooms here in late April." We were not disappointed, as iris, violets, orchids, and "common" trilliums formed a colorful tapestry. We photographed with abandon and kept walking and photographing until early afternoon, but where was our quest species? Had we missed it? Reluctantly, we turned back, but before leaving we paused off-trail by a large moss-covered log for a snack. There, decorating the top of the log, was a diminutive painted trillium. Michael compared it to a young girl's first experience with lipstick—it had a central smear of pinkish red.

In addition to being renowned for its wildflowers, the Smokies are billed as the "salamander capital of the World." Salamanders are called "spring lizards" in the southern Appalachians. Thirty species are found here, including the endemic Jordan's salamander, another desirable quest species. A subsequent visit to the park found us perplexed on locations to hike after a late-April snowstorm blanketed the west slope of the mountains. Also, we had located only a few of the more common salamanders. Consulting the trail guide, we decided to hike Kanati Fork Trail, on the snowless east side. The pleasant and steadily ascending trail led through tall tulip trees, hillsides of wildflowers, and endless birdsongs, but produced no salamanders. At the top of the divide, we had lunch at the base of a red oak and prepared to retrace our steps downward. I turned over a rock near the oak tree on a whim, and underneath it was a Jordan's salamander! The species is found only in wooded habitats at high elevations and is known as a "cool- and wet-weather-loving species." I was elated, as here was another quest completed, and the three miles down seemed to fly by! To top off the day, what was at the bottom of the ridge but a blooming painted trillium—once again, decorating the end of the trail.

BEAVER SIGN

SUSAN POST

It was gray and raining—a duck-friendly kind of day—during a March visit to the Cache River State Natural Area in southern Illinois. There were no wildflowers to see, and the birds were hidden and silent. Yet there was still a lot to observe, and I noticed a surfeit of beaver activity. Quite a few small cypress saplings had been cut on the shore, and floating in Heron Pond, I noticed, were many debarked twigs and branches. Several trees supported scars from gnawing, and a few new raised mounds had appeared on the edge of the swamp.

A swamp-side tree has been cut and skinned by an ambitious, hungry beaver (*top*). A beaver-girdled tree will soon die (*bottom*). Dry beaver scat resembles compact balls of sawdust (*inset*).

Back in Champaign, curious about beavers, I visited our library and also spoke with Illinois Natural History Survey mammalogist Dr. Joseph Merritt. I found that a beaver's habitat must include a year-round supply of water for swimming and diving, floating logs, and underwater burrow entrances for safety from predators. While underwater, a nictitating membrane covers and protects each eye of the beaver. This membrane is a transparent inner eyelid that permits a clear field of vision. A beaver's ears and nose can be closed voluntarily and its lips sealed behind bright-orange incisors to allow chewing underwater. Adapted for gnawing, the four incisors are formidable chisels that are always growing. Beavers live a crepuscular and nocturnal life. They have small eyes and ears (poor eyesight and hearing), but an extremely sensitive nose, allowing them to smell their food before they choose it. Perhaps this acute sense of smell explains why no matter how silent I am, I have yet to sneak up on a beaver at dusk. Beavers need a mixed diet and consume bark, leaves, twigs, and roots of woody plants growing near the water. Fast-growing tree species, with soft wood that is easy to chew and peel, are preferred as food and building materials.

How do they choose a tree? Observations near wetlands often show trees that have been gnawed, but then left alone. Beavers were assessing their nutritional value. Distance from the water also plays a role. The greater the distance, the more tree selective the animals become. Beavers cut only small specimens of less-preferred trees, while size does not seem to matter for their favorites. Also, often to their detriment, they do not know the direction the tree will fall. For a snack, a beaver strips bark from a stick or branch held in its front paws, creating a feeding bed (many floating debarked sticks) in shallow water. The raised mounds I saw, consisting of a pile of mud and leaves, were scent mounds and serve as a warning to intruders. Here the beaver applies odoriferous anal-gland secretions, and the mound elevates the point of odor release.

On that March day, I was lucky to find the ultimate prize—beaver scat. Reference books, and Dr. Merritt, stated how hard it was to find beaver scat. Beavers defecate in the water, and their scat is a mix of plant fibers and woody material—like sawdust. Usually, their scat quickly breaks down in the water, but the Cache was experiencing a drought, so there, at the side of the swamp, lay this mammalian treasure. My beaver observations had taken on new meaning—they now included more than just visions of gnawed wood, but a small, elegant pile of evidence from a well-fed beaver.

WINTER STONEFLIES

SUSAN POST

The life of fall and winter stoneflies is a notable exception to the rule that insect activity ceases in the depths of cold weather. Theodore Frison, an Illinois stonefly specialist, commented in 1929, "Such phenomena cannot fail to excite wonder and forcefully bring to mind the wonderful adaptiveness of the insect world—one of the chief reasons why insects so outnumber all other kinds of animal life." Twenty-four species of the state's eighty native stoneflies appear as adults from November through March. With a habit of congregating in places exposed to the warming of the sun's rays, they can be found crawling about on exposed tree trunks, fence posts, or rocks located close to a stream. The light-colored concrete bridges, characteristic of Illinois's back roads, are also a beacon of warmth to a stonefly.

A mated pair of winter stoneflies enjoy the sunshine. Note that the male is wingless (*top*). A winter stonefly basks on the relative warmth of a concrete bridge (*bottom*).

Stoneflies belong to the insect order Plecoptera and are considered a "sensitive" order—meaning the insects are restricted to habitats where there is little human development and disturbance. Clear water with high concentrations of dissolved oxygen is necessary for their survival. In appearance stoneflies are about a half inch long, have two pair of wings, and are rather drab in color. The adults, although terrestrial, are seldom found far from water and frequent riparian vegetation or rocks near the stream. They are poor fliers, with crawling the preferred mode of movement. The nymphs are often found under stones in streams, hence the common name of stonefly. Winter stonefly nymphs are chiefly plant feeders, while the adults feed on blue-green algae, or not at all. Mature nymphs emerge from the water, find a suitable perch, and metamorphose into an adult. The adults mate, lay eggs, and die, usually living less than a month. Depending on the species, the entire life cycle may take one to three years, with the majority of the time spent as aquatic larvae.

While I am bundled up in fleece and down to view these insects, the small stoneflies produce compounds—glycerol, proteins, and sugars—that function like antifreeze. Although sunbathing in Illinois from November through March is unappealing to most humans, the bridges over creeks emptying into the Middle Fork of the Vermillion River and the vegetation along clear rocky streams of southern Illinois are perfect places for stoneflies to absorb the fleeting sunlight of winter and for us to observe these most unusual Illinois insects. A visit to the Middle Fork River on January 24, 2010, even inspired this poem in my journal:

Winter Stoneflies above Stony Creek

White concrete bridges,
Beacons of warmth.
Nude sunbathers gather,
Looking for love.
Absorbing weak sunlight,
Coupled in pleasure.

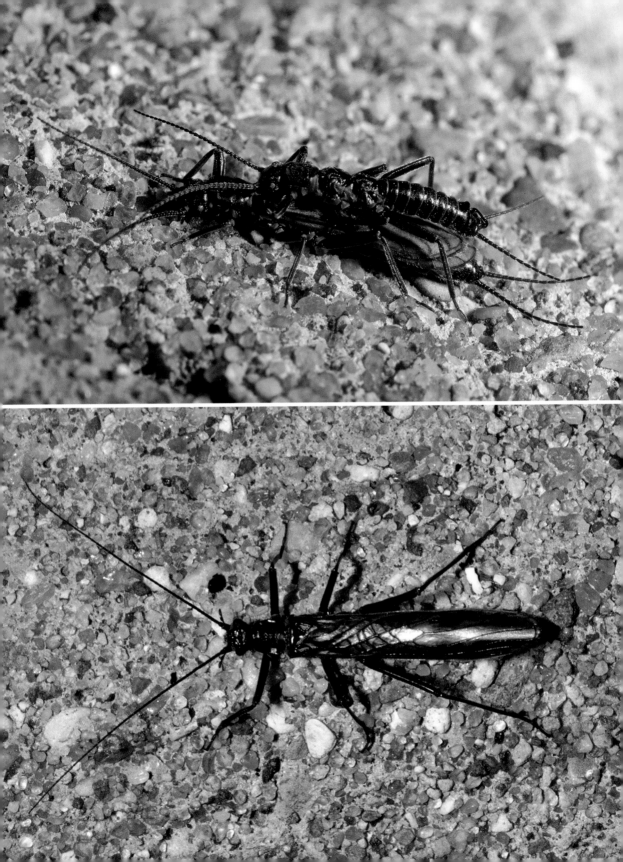

VICTORIA'S SECRET

MICHAEL JEFFORDS

Mention Florida, sun, sand, water, and the colors pink and black, and what inevitably comes to mind? If you are a male, it may be that latest issue of *Sport's Illustrated*—the swimsuit issue—or even the Victoria's Secret catalog, which seems to show up in the mail whether you order from it or not. On a late-spring photo trip to Florida, however, I learned to associate these two colors not with pink flesh and lacy bikinis, but with the delicate pink and white plumage of roseate spoonbills and the stiff Poe-like adornments of black vultures. While the connection between these colors may seem natural, the occurrence of spoonbills and vultures in close proximity in nature is anything but.

Roseate spoonbills have distinctive profiles whether in flight or when feeding. (*top*). Black vultures gather around an American alligator, perhaps waiting for it to make a kill (*middle*). An unusual gathering of black vultures and roseate spoonbills occurred at Myakka River State Park (*bottom*), its purpose unknown.

This unlikely gathering occurred at Myakka River State Park, inland from Sarasota in central Florida. On previous trips to Myakka River, I noticed that the park was a gathering and roosting place for black vultures. In fact, the parking area near the marina is notorious, as untended cars are often stripped of their windshield-wiper rubber by the scores of bored vultures! As a photographer, though, the area is irresistible due to the close approach of many species of wading birds—limpkins, several species of herons and egrets, and, of course, roseate spoonbills.

This latter species is certainly one of the most bizarre and beautiful members of our North American avian fauna. Spoonbills swing their spatula-shaped bills back and forth in shallow water, filtering the bottom mud for their preferred food—crustaceans, aquatic insects, and even frogs and salamanders. Like flamingos, the pink color of the roseate spoonbill comes from pigments contained in the shellfish it ingests. They often feed in large groups, moving through the shallow waters in seemingly choreographed synchrony.

Black vultures, on the other hand, have an entirely different lifestyle and feeding strategy. The black vulture is a scavenger and feeds on carrion (dead creatures), but will also eat eggs and newborn animals. They live in a variety of habitats, including grasslands, savannas, and moist forests and wetlands. At Myakka River, black vultures roost along the river, often high in the trees, waiting for any opportunity to feed. Black vultures have also adapted to human presence and regularly feed at garbage dumps and other areas where food scraps are available. Picnic areas in many parks are often picked clean by scores of these shadowy opportunists.

To a scientist, these two species congregating together in the shallow wetlands—the vultures on the mudflats and the spoonbills feeding eagerly in the adjacent shallow water—presented a puzzling dichotomy. The birds interacted with one another only when one approached the other too closely—vultures would peck at the spoonbills, and the spoonbills would sway their bills back and forth above the water. Otherwise, they seemed content to occupy nearly the same place in space and time.

So what actually is Victoria's secret? It's really quite simple. Never underestimate Mother Nature—her juxtaposition of colors and forms is infinite, variable, seldom predictable, and worthy of any Paris or New York fashion designer.

HIGH SALAMANDER

MICHAEL JEFFORDS

For years Sue and I had longed to visit the Rocky Mountains in the fall to experience the aspen in full color, and one September off we went. The landscape that greeted our arrival at Rocky Mountain National Park inspired a bout of journal writing—the scenery was overwhelming.

It was like driving through mountains of glowing, botanical lava. The bright double-yellow line down the center of the winding, dipping, curving highway paled by comparison. ...I now understand gold fever, the lust that must have driven the prospectors of old to head into the gold fields of the West. In the minds of these men, their destination contained nuggets, veins, even mountains of gold. Many years of brutal toil, rewarded only occasionally by a few "flakes of color," drove many to disappointment and despair. Most eventually packed up and went home. Today, we drove through a landscape that was every prospector's dream, a terrain rich with nuggets, veins, even mountains of a different gold—the aspens are in full color.

An encounter with a tiger salamander, high in the Rocky Mountains, is an unexpected treat.

Despite the hypnotically beautiful drives, we had also come here to hike, and climbing toward Bierstadt Lake, approaching nine thousand feet in elevation, we had a most curious and unexpected encounter. Sunning on a trailside boulder, seemingly enjoying the same view we were, was a rather large, dully colored salamander! The temperature was hovering around 40°F, and winter was not far away. What was this creature doing here? Back at our campsite, we tentatively identified it as a tiger salamander, a familiar species, but those found in Illinois are smaller and usually have bright-yellow markings. This one was a blotchy brown and almost ten inches long—a giant.

Consulting with colleagues when we returned to Illinois yielded a plethora of information, including that tiger salamanders (confirming our identification) are rarely seen in the open. They live in burrows down to two feet below the surface. Most eat insects, worms, and other invertebrates, but large examples can also consume small frogs and even baby mice. They are found in a wide variety of habitats, even thriving in the desert Southwest and parts of Canada. Large individuals lose their golden color and are said to be very fast and seldom miss prey items they pursue. This creature, however, seemed content to bask in the warm sunshine and ignore us. A map of their distribution shows they occupy the entirety of the central United States, from the Rocky Mountains to the Mississippi River basin, but they also range into Wisconsin, Michigan, Indiana, and Illinois. Nowhere, however, could we find any information about their occupying the higher elevations of the Rocky Mountains. Tiger salamanders live for up to fifteen years in the wild, and the size of this individual indicated that it was thriving. After observing and photographing this unique creature, we moved on, leaving *Ambystoma tigrinum* to enjoy the view.

HOATZIN LINEUP

MICHAEL JEFFORDS

The phrase *living fossil* is not a scientific term, yet it is often overused by the news media regarding oddities of nature. The term means "a living species of organism that resembles a species known only from fossils." Examples include the coelocanth, a fish thought to be extinct but discovered in 1938; the horseshoe crab, a species little changed over the past 400 million years; and the hoatzin, a bird found along tropical streams in South America. Hoatzins are primitive birds that retain claws on the wings during their juvenile stages. They are often linked with the famous feathered dinosaur *Archaeopteryx* that was discovered in Germany in 1860. *Archaeopteryx* lived around 150 million years ago and was once thought to be the "first bird." Recently discovered fossils in China, however, have displaced it.

Four, then five hoatzins line up on a branch after a busy bout of leaf feeding. They seemed to like to be together, and this branch ultimately supported six hoatzins. Their resemblance to artist's renditions of *Archaeopteryx* is quite startling, giving credence to the term *living fossil.*

During a birding expedition to the remote Madre de Dios River in Amazonian Peru, the unique hoatzin was at the top of my "must-see" list, although I was skeptical about my chances. During the trip, the group would traipse through the rain forest, seeking elusive quarry, such as antbirds, antpittas, and tapaculos, often diligently peering into tangled thickets for long periods. I would soon grow tired of this sedentary quest and wander off alone. One afternoon I happened upon a small oxbow lake, not far from the lodge, and this became my favorite place to walk. The trail was rife with long-wing and owl butterflies, fascinating flowers, and, on one special afternoon, hoatzins! Not just one, but a lot of them. About the size of a chicken, several of these birds were noisily feeding in the vegetation that lined the lake.

Hoatzins are the only birds with a digestive system similar to a cow's—they eat only vegetation, which is processed by grinding in a large crop (not the familiar gizzard found in other birds). They use fermentation to digest the leafy material, and hoatzins have a very "unusual" smell. First described in 1776, the hoatzin is thought to be related to cuckoos, and fossil evidence from France suggests they may have been around since the Eocene, more than 36 million years ago. Hoatzin young are fed a leaf paste regurgitated from the crop and are excellent swimmers. They usually drop into the water to avoid predators. Adults are clumsy fliers, often falling into the water and noisily crawling back up into the foliage, where they sit with the wings outspread to dry. It was this activity that drew my attention. I sat quietly on the opposite shore and watched the show. Several individuals were active, and I could hear a variety of hisses, yelps, and hoots, but saw only fleeting glimpses of the birds. As I was about to leave, a single bird flew to a large bare branch, in clear view, and rocked back and forth to stabilize itself. I was ecstatic and began photographing. In short order, the single hoatzin was joined by a companion, then another, until there were six lined up before me. They rocked and squawked for quite some time as I clicked away and eventually got my fill of living-fossil photos.

DOWN THE HATCH

MICHAEL JEFFORDS

As biologists, we often go to popular parks and natural areas, but at very odd times. We do this for several reasons: to avoid the copious crowds of the "high" seasons, to satisfy the need to just get away from work for a while, and, perhaps most important of all, in these "off times" we frequently observe things that would go unnoticed in a park busy with tourists. Such was the case when we explored remote Pickett State Park in East Tennessee during a cold, but snowless, December. Pickett State Park is noted for its spring wildflower display (April) and its incredibly scenic swimming and fishing lake (summer)—one of the most beautiful man-made lakes I have ever seen. With the park's massive stone arches that rainbow across the landscape, it is also the destination of many a fall hiker. But we were here to relax and enjoy the wintry solitude. We had low expectations for wildlife and expected to enjoy only frozen waterfalls, massive shelter bluffs, and the sandstone arches, silhouetted against a cold, gray winter sky. We saw all of these things, but were surprised by unexpectedly rich wildlife experiences. Pileated woodpeckers hammered diligently just outside our cabin window, their bright-red crests breaking the umber monotony of the wintry woods. Along the lakeshore, abundant evidence of fresh beaver activity included many downed trees and others that had been trimmed and dragged into the cool blue water to soak. Beavers feed on tree bark during the winter. We also witnessed a pair of river otters in a quiet alcove of the lake. We sat quietly and watched from the shoreline as they cavorted, splashed, and swam about, oblivious to our presence.

The highlight of our sojourn was a glimpse into the lake's food chain in action. As biologists, we know that it's an animal-eat-animal world out there. But to witness it firsthand is something very special. On a morning hike over a natural arch that crosses the lake, we noticed a small pied-bill grebe swimming slow circles in the deep waters of the lake. It paused, dove quickly into the water, and disappeared from view. Perhaps twenty seconds later, the grebe reappeared with a fish that was nearly as long as its body. We both commented that this was certainly a case of "biting off more than you can chew!" Undeterred, the grebe spent a few seconds wrangling the wriggling fish and finally grasped it by the head. Swallowing it any other way would result in the sharp dorsal fin puncturing the esophagus, maybe even killing the bird. The grebe rested a few seconds and, in a series of gulps, amazingly swallowed the entire length of the fish. Why have many meals of small fish when one giant catch can do the trick? Most field guides state the pied-billed grebe catches, kills, and eats amphibians, crayfish, other arthropods (insects and so forth), and small fish. Obviously, our grebe had higher ambitions and had failed to read the textbook.

A pied-bill grebe captures a rather large fish (*top*), positions the catch head first for swallowing (*middle*), and prepares to swallow the massive meal in a series of gulps (*bottom*).

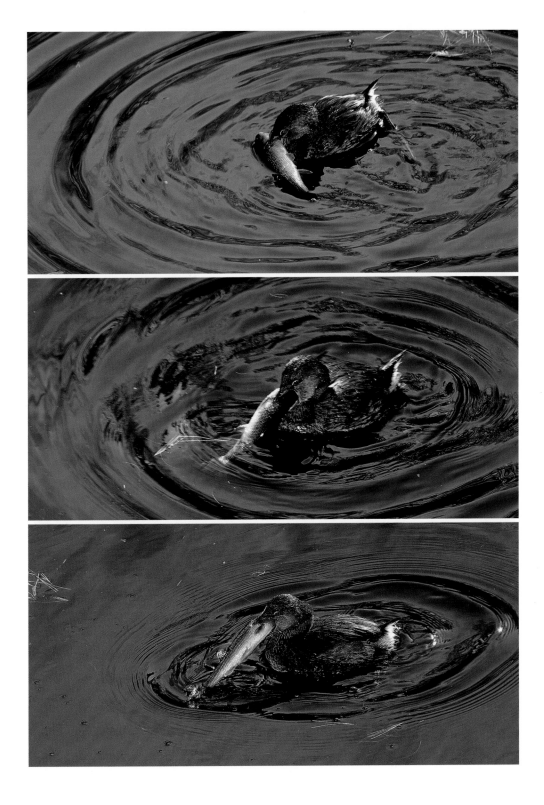

DEMONS IN THE DUST

SUSAN POST

Have you ever cringed as the heroine in a B horror movie opens the door to a sinister knock? Her insatiable curiosity almost always ends in doom. Not so for the curious behavior of scientists, as we are rewarded for peeking into places no one else has been. For example, consider the humble ant lion. They are insects that belong to the order Neu-

An ant lion adult (*above*) provides no clue to its voracious upbringing. This ant lion larva is safely ensconced at the bottom of its carefully constructed trap, waiting for prey (*top left*). An ant lion larva is on display away from its pit (*top right*). Ant lion larval pits (*bottom*) occur in sandy or dusty soil.

roptera and are related to lacewings. Neuropterans have complete metamorphosis—egg, larva, pupa, and adult. Ant lion adults have been described as "slender insects of a delicate nature." They have lacy net-veined wings, a slender body, and conspicuous knobs at the ends of their antennae. An adult ant lion's life is measured in days, as its sole purpose is reproduction. While the adult is a creature with a fragile, quiet nature, these terms would not describe an ant lion larva. Most larvae are less than a half inch long, flat, with short, stiff hairs fanning out along their sides. They have broad heads armed with sickle-shaped jaws that are spiny and bristled near the base and smooth, sharp, and curved at the tips. Ant lions have specialized jointed legs: the two front pairs are small and spiny, while the hind legs are strong and designed to anchor the larva in its home. Their life span can be anywhere from one to three years, depending on food supply. Ant lion larvae

make their living in the bottom of small conical pits. We have all seen them in the sand and dust. In rural areas, these insects are known as "doodlebugs." Once the ant lion larva finds a suitable place to dig, it moves in a series of concentric spirals, each deeper than the last, until the pit is excavated—a funnel-shaped sand or dust trap. The most successful ant lions have traps in places protected from wind and rain—everywhere from deep sheltered bluffs to the floors of ancient barns. Each trap is a simple steep-sided pit. Two inches across is considered huge. Depth varies in proportion to the width of the opening, and the sides typically have about a forty-five-degree slope. Ant lion larvae seem to have an engineer's sense of slope stability, as they build their traps as steep as the soil will tolerate. To excavate the traps, ant lion larvae use their broad shovel-like heads to flip sand out of the pit and over the rim. Once completed, the ant lion waits patiently at the bottom. Ant lion larvae will lie in wait for days, even weeks, for a meal to stumble into their lairs. As they wait, they tend the trap, periodically turning and evenly distributing the sand so there are no telltale sand piles next to the pit. The pit is created so that when prey, usually ants, wander into it, the extra weight causes the sand to collapse. The ant falls toward the bottom of the pit, where the ant lion's open jaws await. The ant lion will throw or kick sand out of the bottom of the pit to further destabilize the walls, creating mini landslides that hasten the descent of the prey. Prey are pierced, paralyzed, and their fluids "sucked out." The ant lion's sickle-shaped jaws house canals that transport blood and bodily fluids from its prey. The desiccated carcass is then, unceremoniously, tossed over the rim of the pit.

THE LAST OTTOE?

MICHAEL JEFFORDS WITH SUSAN POST

One of the most poignant encounters we have experienced came during a visit to the Cincinnati Zoo. We were there to see the world-renowned live-insect exhibits, but had arrived early and were wandering the grounds. Sue and I happened upon a life-size bronze statue in a small courtyard of Martha, a passenger pigeon. This humble pigeon was the last

James Wiker views perhaps one of the last examples of the rare Ottoe skipper on a remote Illinois hill prairie (*above*). Will the species be seen again in Illinois? An Ottoe skipper nectars on a coneflower (*bottom*).

example of a species that had once numbered in the billions and darkened the skies of eastern North America. When Martha quietly slipped from her perch on September 1, 1914, the passenger pigeon became extinct. While we had not witnessed the final demise of this species, the point was made, and we were moved and saddened by this encounter.

Several years later, we were teaching a field class for thirty-plus people in west-central Illinois on skippers (order Lepidoptera, family Hesperiidae). These small, often drab brown creatures are considered to be butterflies, although less well known than their more colorful counterparts and difficult to identify and photograph in the field. We had achieved some success finding various species, but our last day was to be spent on Revis Hill Prairie Nature Preserve. Hill prairies are dry grasslands that grace the tops of bluffs, usually overlooking river valleys. They are strenuous to access, but the plants and views from the top are worth the effort. We had managed

to coax—and drag—most of our group up the steep two-hundred-foot slope in hopes of seeing the rare Ottoe skipper. This would be a highlight for the class. The only member of our group that did not make the trek to the top was Dr. James Sternburg, our course instructor and a coauthor of *A Field Guide to the Skipper Butterflies of Illinois*. The day was hot, Jim was ninety years old, and he had never seen a living Ottoe skipper. Our coinstructor and skipper expert, James Wiker, had encountered Ottoes here many times in the past, but stated they had declined drastically in the past decade. We searched, and searched, and finally, as we were preparing to leave, Wiker found a lone Ottoe skipper. Everyone was ecstatic at seeing this rarity in all its drab glory—everyone except Dr. Sternburg, who waited quietly in a lawn chair far below. Dr. Sternburg was my Ph.D. advisor, and he and I had spent many years as mentor and mentee. I could not let him miss seeing one of the few species of butterfly he had missed in his long lifetime. We caught the skipper and placed it in a small jar, and I headed down the steep path with the prize. Dr. Sternburg was ecstatic to finally see an Ottoe, and he diligently photographed it through the plastic jar. Not ideal, but I wanted to make sure it was returned safely to its hill-prairie habitat. Back up the hill with the Ottoe, I gently placed it on a coneflower so the class could photograph the species. The course ended with everyone excited and happy.

Since then, Wiker and others who search for and study skippers have returned to Revis many times to look for Ottoes. The skipper has been seen only once since then. Our minds could not help but remember the encounter with the bronze Martha. Had we had our own experience with extinction, and was the Ottoe now just a sad memory in Illinois? Only time will tell.

GULLS AT PLAY

MICHAEL JEFFORDS

Animals play, especially when they are young. We have all witnessed the antics of a new puppy or kitten and may even have seen young wild animals cavort about in what we interpret as "play." In most instances, these playtimes involve skill-building activities that prepare the young to fend for themselves. I have watched brown bear cubs in Alaska play at fishing with the mother bear and witnessed the puppylike behavior of spotted hyena pups in South Africa. This play has real-life consequences, as these animals either learn adult skills very quickly or do not survive. It is serious business.

A panorama shows gulls floating downstream and then flying back (*top*). Gulls rest alongside a spillway stream (*bottom left*). Gulls enter white water downstream from the spillway (*bottom right*).

One December at the Rend Lake Dam spillway in southern Illinois, I witnessed what truly must be considered "play" among a large gathering of ring-billed gulls. The water from the large impoundment rushed over the spillway and splashed down the channel in a cascade worthy of Class 4 rapids. Many of the thousands of gulls clustered along the rocky shore simultaneously took to the air, flew upstream, and entered the water just below the spillway in a cacophony of sound. The raucous flotilla chattered their way down the roiling stream for a quarter mile until they reached the stretch of stream where the water calmed. They then took off and flew back upstream to repeat the entire process! I observed this behavior for more than an hour with no apparent diminution of enthusiasm among the participants. Initially, I thought I was witnessing a feeding frenzy, but close observation showed no indication of feeding by the gulls. They just seemed to enjoy the water-park-like ride. A Corps of Engineers employee happened by, and I questioned him on this odd behavior. He said the gulls had been so engaged for a couple of days, and he had not observed such raucous activity before.

While it is difficult to interpret unusual bird behaviors, and we humans tend to anthropomorphize animal activity, I could see no other reason for this behavior other than the excitement of the white-water ride. After several passes up and down the stream, participating gulls would rest on the shoreline, only to be replaced by other gulls on the aquatic roller coaster. I queried several of my ornithological colleagues at the University of Illinois, and they agreed that gulls often play, but none had witnessed this particular phenomenon. A search of various books and websites yielded information on gulls that emphasized their intelligence and cleverness. For example, some species are known to stamp their feet to simulate rainfall to entice earthworms to the surface. It has also been observed that gulls learn and remember new activities, passing on these learned behaviors to their offspring and to other gulls. Over the past several years, I have visited this site at the same time and under similar conditions, but have observed this unusual play behavior on only that single occasion. It's difficult to imagine that this was a unique occurrence. But for the moment, I can think of no other reason for riding the wild waves other than the sheer joy of "being a ring-billed gull."

CHITINOUS SNACK

MICHAEL JEFFORDS

The American alligator is notoriously catholic in its feeding habits, and while considered an apex predator (one that resides at the top of the food chain), it is also an opportunist. Alligators are known to feed on bobcats, adult deer, the introduced Burmese python, and even various types of fruit. They have even been reported to be a "tool-using" reptile: alligators have

been observed balancing sticks and branches on their heads to lure large birds (egrets and herons) seeking nesting materials close enough so they can grab them! Given the adaptability of the American alligator, it should come as no surprise to find them feeding on just about anything, however

An American alligator crushes a large horseshoe crab in its jaws in preparation for ingesting this prickly, chitinous meal (*right*).

unappetizing it may appear to us.

I witnessed an unlikely alligator meal on Sanibel Island, Florida, in the 1980s. This was before much of the island became developed, and there were still miles of relatively untouched beach to roam and explore. Late one afternoon, while beachcombing and looking for seashells, I saw a distant alligator on the edge of a swamp. As I cautiously approached, I heard a tremendous crack, as if someone had stomped a large Styrofoam cooler, shattering it into pieces. The area I was walking through was a narrow beach that fronted a brackish marsh, bordering a mangrove swamp. A ten-foot American alligator had captured an especially large horseshoe crab (at least a foot across). These flat,

spiny relatives of spiders are considered living fossils; fossil examples that are 450 million years old are not that different from living horseshoe crabs. They are marine arthropods, and the entire body is protected by a hard carapace, or shell. In technical terms, they have a very tough exoskeleton, largely composed of chitin. Chitin is a long-chained polymer, a derivative of glucose, but in horseshoe crabs, and other arthropods, it is combined with other materials to form a tough, strong exoskeleton that serves to protect the animal. In fact, few organisms can digest arthropod chitin. The horseshoe crab is really vulnerable only when it is turned over, revealing its relatively tender underparts, and they are often attacked and fed upon by gulls. But, returning to our beach scenario, it would seem that none of the above was relevant to a hungry alligator, as I watched it grab the large horseshoe crab. With the crab situated flat inside its mouth, obviously not the best way to swallow this prickly meal, the alligator had opened its jaws and jerked its head up and down several times until the horseshoe crab came to rest perpendicular in its jaws. Once positioned, the gator demonstrated its legendary bite force (three thousand pounds per square inch) and simply crushed the horseshoe crab. That was the noise that mimicked a cooler being demolished. I watched in awe as it opened its jaws repeatedly to further crush the crab into a more manageable size and promptly swallowed the entire spiny snack. Given the exoskeleton-to-meat ratio of a horseshoe crab, I can't imagine the gator got much nutritional value from its effort, but it surely did solidify its moniker as "apex predator" for this observer!

SNAKE DANCE

MICHAEL JEFFORDS WITH SUSAN POST

Ritualized combat occurs on a regular basis in human society—tune in to a WWE (World Wrestling Entertainment) bout or buy a ticket for a local match to see testosterone-laced males engage in "mock" combat. But what about combat in nature—do animals engage is these stylized rituals? The answer is yes, as Sue and I witnessed such an engagement, early one summer, between large male cottonmouth snakes at Mermet Lake Conservation Area, deep in southern Illinois. We often drive the five-mile perimeter of this impoundment, searching for birds, so we were surprised to find a pair of cottonmouth males engaging in what looked like "endless thumb wrestling" in a small pool along the levy. Although the species is venomous, during the forty-five minutes we watched the bout, neither male opened his mouth or attempted to bite the other. As Sue wrote in her journal, "They were like miniature cobras, writhing, pushing, splashing—a continuous, sinuous entanglement." As with most animal behaviors, it is reported in the scientific literature to "regularly occur," but seeing it firsthand in the wild was quite rare and something to behold.

Cottonmouths, a southern species that inhabits swamps and lowland forests, mate in early summer, and the combat that we witnessed was a competition

Equally matched male cottonmouths engage in a bout of ritualized combat for the right to mate with a female. We watched the event for nearly an hour before moving on.

for dominance. The prize to the victor was the privilege of mating with a nearby sexually receptive female. The technical term is *sexual selection*, defined as "natural selection arising through preference by one sex for certain characteristics in individuals of the other sex." In this case, the male with the greatest endurance likely received the "right" to mate with a female cottonmouth. The losing male, tired and defeated, likely survived to grow for another season to perhaps have his chance again next year.

On our frequent visits to this site, we seldom encounter cottonmouths and wondered why this year was so different. Mermet Lake, despite being a haven for wildlife, is a highly managed system. In summer large diesel-powered pumps spew water from the surrounding wetlands into the body of the lake for fishing. In fall the process is reversed, and lake water is removed to flood the surrounding wetlands for waterfowl hunting. On this particular day in June, the summer pumping had just been completed. The normally wet canal, from where the levy dirt had been removed when the lake was impounded in the 1930s, was reduced to a series of small pools. We speculate that the reduction of the moat to mere pools had brought these two male cottonmouths in close proximity to one another. We happened along at just the right moment. The only "ticket" we needed to purchase for this remarkable WWE event, conducted in the shallow watery ring of a southern Illinois swamp, was the cost of the trip from our home in central Illinois. It was more than worth the price of admission!

FROM ABOVE

MICHAEL JEFFORDS

Imagine yourself as the lead goose in an asymmetrical V, flying high over the Illinois countryside. What a view that must be! As it so happens, over the span of several years, I've had the privilege of being that "lead goose." I have spent considerable time soaring over a large portion of Illinois, not on my own power, but safely ensconced in a small plane, piloted by friend and colleague Dr. Phil Mankin, a University of Illinois wildlife ecologist. He and I conceived of these flights as an informal project to see Illinois from a different perspective. The photographic survey was neither systematic nor complete, but it allowed us to pick and choose interesting aspects of the landscape to depict.

The confluence of the White River (*lower*) and the Wabash River confirms that the White, flowing from Indiana, has lost much of its riparian vegetation, as it is carrying an enormous silt load (*top*). The Great S-Bend of the Mackinaw River showcases the fact that rivers, in their natural condition, meander across the landscape (*bottom*).

Viewing Illinois from above, one cannot help but be struck by the dramatic revelation that Illinois is a profoundly human-dominated landscape. Illinois is a prime example of a region that has left the Holocene (the currently accepted geological epoch) and proceeded into the Anthropocene (viewed by some scientists as the epoch during which human activity is the dominant influence on climate and the environment). Virtually all of my aerial images of Illinois depicted evidence of human interaction or activity with the landscape. There were really no expanses of undeveloped or undisturbed land large enough, when viewed from the air, to fill an entire image. Everywhere I pointed the camera, human intrusions entered my viewfinder! Even in deepest southern Illinois, flying over the Shawnee National Forest, the human footprint was evident: power-line pathways cut through the forest, square patches of planted nonnative evergreens appeared in an otherwise rugged landscape featuring mostly oak-hickory forests, and roadways, from gravel to interstate highways, crisscrossed the land in a geometric grid. Later, I learned that no point of land in Illinois is farther than eight hundred yards from some sort of roadway.

Flying up and down the rivers also provided quite a revelation. Most streams in Illinois have been "improved" over the years—either by channelizing and straightening or by damming. The proximity of vast tracts of agricultural land next to the free-flowing Wabash clearly marks it as a river of the Anthropocene. One startling view was where the White River entered the Wabash in southern Illinois. The Indiana stream was almost white with silt—the best soil Indiana had to offer! For much of its length, the White flows through agricultural lands with little or no riparian vegetation to catch silt-laden runoff. Flying over the Great S-Bend of the Mackinaw River in central Illinois also offered practical instruction into *fluvial geomorphology*—rivers meander, or more succinctly, they go where they want!

What was the upshot of all this random visual eavesdropping from the Illinois sky? Simply put, you can't hide anything from the sky.

TURKEY TANGO

SUSAN POST

No visit to southern Illinois is complete without a stop at Mermet Lake Conservation Area, near Metropolis. During any season, there is always something to see. Yet during one late October visit, I commented to Michael that this just might be the first time we had not seen some form of wildlife activity. As soon as I uttered those words, I noticed a shape approaching from the distance. The day was windy and wet, and my first thought was a stray garbage can was rolling down the road. As we drove closer, the black and white blob resolved into a pair of yearling turkeys (called jakes) involved in a tussle. First they wove to one side of the road and then to the other. But when they hit the pavement, it was as if they were in a dance competition, and it was their turn in the spotlight. Suddenly, I could not help but recall the often incomprehensible terms Miss V, my University of Illinois ballroom dance instructor, would shout out during class. Here, the mysterious terms were brilliantly illustrated by this turkey pair. They were performing the tango—a dance of close embrace. There was no space between their bodies. The *abrazo*, or tango hug, was performed at chest level and with turkey heads—neck-on-neck and beak-to-beak. The *gancho*, usually executed by hooking one leg around a partner's leg or body, also

This pair of young turkeys spent more than forty-five minutes engaged in a seemingly orchestrated "dance for dominance." We witnessed the beginning of the dance, but left before any hierarchy could be established.

involved their brightly colored necks. A male first hooked left around the opponent and then right; this was an intricate interweaving of long turkey necks. Deeply interlocked beaks bound the pair together. The claws on their toes scruffed the pavement, sounding like taps on shoes, but never missing a step—forward and back, back and forward. I saw no *arrepentidia* (evasive actions that allow a dancing couple to back away to avoid dance-floor collisions), as these jakes were not backing down, and they were oblivious to our close approach. We were spectators to their performance for more than forty-five minutes, yet never did they *parada* (stop). When we finally drove off, the jakes were still locked in a sinuous embrace.

To learn more about "dancing turkeys," I consulted a National Wild Turkey Federation publication, *The Wild Turkey*. By late fall, this year's pullets (young) have matured and separate into male and female groups. In male flocks, a "pecking order" is established by fighting. What we had witnessed was a pair of sibling males fighting for dominance. This was probably the first of several "dances," as later on many such contests would follow in the name of "keeping order" in large flocks.

Anibal Troilo, an orchestra director wrote, "The tango is discovered little by little, and it chooses us. When it does, it gives us a glimpse, but it remains, as it has forever, surrounded by a halo of impregnable mystery." His words seem perfectly suited for our visits to southern Illinois. We visit often and feel like we know the place, but in reality it takes only a pair of wild turkeys to show us that we really don't have a clue.

TRULY BIZARRE

The World Wide Web is rife with videos of bizarre happenings and seemingly unique occurrences that have gone viral and entertain millions of viewers. Some are genuine, while others have been manipulated through the miracle of digital technology. It can be difficult to tell truth from fiction in the modern age.

On a trip to South Africa, we visited the Cape of Good Hope. As we returned to our car, we had a visitor who was feasting on our bird seed–filled bags used to steady our cameras. We had failed to close our car window (*previous page*). But you can't say we weren't warned (*left*)!

Cable television takes full advantage of unusual recorded events with innumerable shows with the titles *World's Dumbest, Craziest, Smartest...* etc. They amuse us by showing how easily our neat, trim, organized human society can be subject to circumstances simply beyond our control. In nature it's often the case that events seem to proceed by way of a given set of ecological "rules," which have evolved over the millennia. These include trophic structure and the food web, biogeography, community structure, and microbial ecology. While these topics don't exactly make for scintillating conversation on a daily basis, they are vital to the overall functioning of our planet.

Somewhat closer to home, we can all anticipate the consequences of whacking a hornet's nest with a stick, or picking up a bumble bee with a bare hand. In both instances, the result is a given—we will be stung. Venturing into an Illinois blackberry patch in summer to reap the harvest will likely yield a plethora of chigger bites and the more than occasional tick. Choosing to picnic in a patch of poison ivy would certainly be a rash endeavor.

We like our world to be predictable, stable, and familiar. However, in many instances, chaos theory—events set in motion that cannot be predicted, although produced by a deterministic system—intervenes and produces downright bizarre interactions. Over our lifetime as field biologists, we have witnessed events that we could never have anticipated. Who could have guessed that a children's playground would become a stag beetle proving ground, or that a citizen could actually think they had found a living trilobite, or that carp watching would become a contact sport?

One thing that sets the biological sciences apart is that we deal with constantly evolving, living systems, populated by creatures that may react in unexpected, even bizarre, ways. Add to this mix the complex boundary that modern society has erected between the natural and managed worlds, and you have the potential for endless combinations of unexpected events. This chapter depicts only a sampling of the bizarre observations and encounters we have experienced. They have helped sharpen our senses and hone our analytical skills so that we never take anything for granted when exploring nature and the natural world.

BEETLE PLAYGROUND

MICHAEL JEFFORDS

Turning your tax dollars into "beetle biomass" is not likely on anyone's "money well spent" list. I didn't know it was possible, until a July phone call to my office. "You've got to get over to so-and-so park [name withheld to protect the insects]; there are hundreds of stag beetles!" Really? I thought. How could this be? The only species seen in central Illinois

is the reddish-brown stag beetle (*Lucanus capreolus*). Occasionally, these large insects are observed at porch lights or lying crushed on sidewalks around town. Known locally as "pinching bugs," they are unmistakable. The large, curved mandibles of the males are their most distinguishing feature. These are woodland creatures whose larvae live in decaying wood—downed trees and tree stumps—and take two years to develop. Over the course of my entomological career, I had probably seen a few dozen of these beetles. But hundreds in a city park had to be a rampant exaggeration.

My curiosity more than piqued, Sue and I headed over to the very small city park in question. Its most notable feature was a large playground, full of colorful play equipment and a group of small children enjoying the exercise. From a distance, nothing seemed unusual. As I approached the playground, I noticed

Male stag beetles cluster on a branch and joust with each other for dominance. The "lesser" males are often dislodged and fall to the ground. Females wait nearby and choose their mate from the most powerful males—a classic case of sexual selection (top and bottom left). A group of stag beetles clusters at the base of a nearby tree. Perhaps they were not used to such large numbers of their kind and were as puzzled and astonished as the entomologists who found them (bottom right).

movement in the branches of a nearby tree that turned out to be stag beetles. At least a dozen of them were lined up; the males were "facing off" against each other in a show of force. Male stag beetles use their large mandibles to joust with each other to establish dominance and access to the nearby females (with smaller mandibles)—a classic case of sexual selection. While the gathering was quite impressive, where were the "hundreds" I had been promised?

A quick walk around the playground revealed that the park district had been using hardwood mulch as a playground surface for many years. I dug around and found that the mulch was many inches (perhaps feet) deep, and the accumulation had decayed into a massive stag beetle hatchery! Unwittingly, park staff had created the perfect habitat for these uncommon insects. Indeed, there were hundreds of the beetles around the playground, many just emerging from the decaying wood or resting at the bases of nearby trees. The children noticed our presence, and their teacher asked what we were doing. When we pointed out the beetles, she called the children over and said, "You listen to these folks. They know all about the pinching bugs!" We had a rapt audience, at least for a short time.

Later that day, I returned with several interested colleagues. We were busy photographing this bizarre phenomenon when a bicyclist whizzed by on the nearby street, noticed our presence, and screeched to a halt. "What are you guys doing?" We explained that we were entomologists from the University of Illinois and that we were photographing this most unusual gathering of stag beetles. He look puzzled for a moment, before saying, "You know, I've lived in this university town all my life and knew there were people like you around. I just never expected to meet any of them." With that he rode off, and we continued to enjoy the mass emergence of *Lucanus capreolus*.

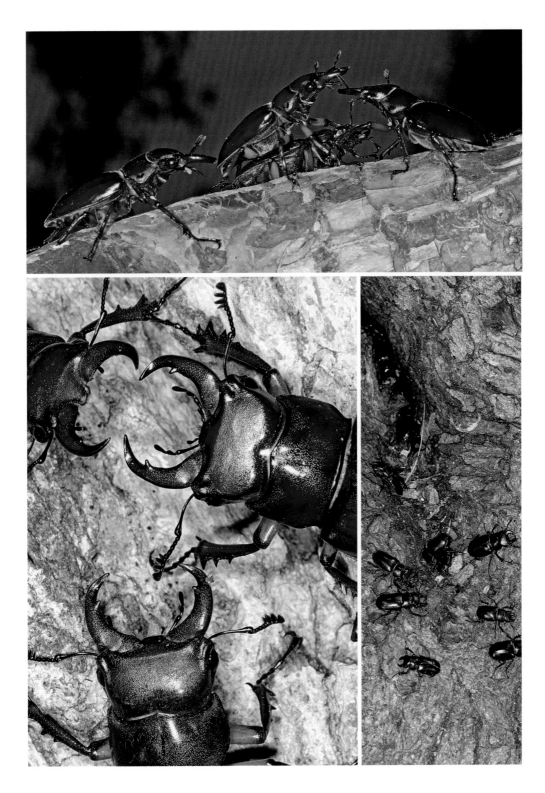

TRILOBITES TODAY?

MICHAEL JEFFORDS

Over the years, I have given seemingly innumerable talks to various groups around the United States, and some of my most interesting moments have come from the question-and-answer sessions after each presentation. As they say, "No question is inappropriate, if asked and answered with sincerity and a desire to advance knowledge." With the advent of the World Wide Web, my world has expanded, and I am a member of various forums composed of scientists willing to answer questions from the public. Most are relatively easy to address: What is it, or what is it doing? But one question I found so bizarre, it immediately made me chuckle. I was amused until I looked at the question through the eyes of a nonscientist. The question was a simple one. "I found this in my backyard, and it looks like a trilobite. Is it?" My first reaction was *Really?* But when I looked at the submitted image, I had to admit, it did superficially resemble a trilobite. However, the likelihood that a living example would appear in a southern Illinois backyard in the twenty-first century is precisely zero. I thought that was common knowledge, but apparently not. The item mistaken for a trilobite was the egg case of a Carolina praying mantis. In all fairness, both the egg case and the trilobite seem to

A fossilized trilobite, *Flexicalymene celebra* (*above*) has a distinctive three-lobed structure.

A Carolina praying mantis egg case (*right*) superficially resembles a fossil trilobite. An adult tadpole shrimp's habitat is an ephemeral pool in a depression in solid sandstone (*far right*).

have a similar three-lobed structure, and if you were unfamiliar with the groups, perhaps I could see the genesis of the question. The inquiry helped me realize that when interacting with the public, I should take nothing for granted.

This experience started me thinking about our perceptions of things and the role that background knowledge plays in our view of the world. I couldn't help but wonder what other animals of today might be mistaken for trilobites. Horseshoe crabs came to mind, but perhaps one of the best examples was an organism I encountered in the ephemeral rock pools of Canyonlands National Park in Utah—the tadpole shrimp. Considered a living fossil, little changed over the millennia of its existence, the crustacean has its own order (Notostraca) and a single family (Triopsidae). Tadpole shrimp spend the dry season, when there is no water in their pools, safely ensconced as eggs in the accumulated silt and debris. When rain fills the rocky pools, they hatch, feed on a variety of organisms and organic debris, and eventually become adults. For their life cycle to be completed, the pool must retain water for around three months. Adults can reproduce sexually (rarely) or parthenogenetically (without sex from unfertilized eggs), and some populations are even hermaphroditic (individuals contain both sex organs and fertilize themselves). If a species lives in such a tenuous habitat, it pays to have a number of strategies! Coincidentally, I had encountered these creatures as a child. They were (and still are) advertised in comic books as "sea monkeys" and sold as novelty aquarium organisms. So always be wary of question-and-answer sessions, and remember to be nice. Trilobites indeed!

DID I JUST SEE THAT?

SUSAN POST

Everybody desires to observe interesting things. In nature it can range from a new bird in your backyard to a wild animal seen while on vacation. When I travel, I research sites extensively to increase my probability of seeing as much animal and plant diversity as possible. I also try to go at optimum viewing times—the Great Smoky Mountains in April for wildflowers, early winter in New Mexico to see sandhill cranes, and early summer in central Illinois for butterflies. As a birder, I am constantly stalking rare-bird reporting websites and reading past trip summaries from the various birding tour companies. As Louis Pasteur said, "Chance favors the prepared mind."

A cartoon depiction of an unlikely series of events in the Everglades will have to substitute for reality. We had failed to carry our cameras, thinking there would be nothing to see (cartoon by Joseph Spencer)!

A February trip to Everglades National Park had me dreaming about flocks of roseate spoonbills, nearby anhingas drying their wings in the mangroves, and views of the elusive short-tailed hawk. The latter (a new bird for my life list) was almost "guaranteed." A colleague had visited the area two months earlier and told me in great detail where to go and at what time I needed to be there. Daily during our stay at the tip of the Florida peninsula, I went to the appointed area and searched—up, down, and all around. No short-tailed hawk testing the thermals ever came into view—so much for preparation.

One afternoon Michael and I decided to hike the Bear Lake Trail, a path through hardwood hummocks and thick mangrove forests. The waters of Bear Lake were steeped with tannins and tea colored. Mangrove roots were interwoven in a chaotic maze, trying to maintain the seemingly precarious perches of the tall trees that ringed the shoreline. I carried only my binoculars, thinking I wouldn't be seeing anything worth photographing. Sightings of several overwintering warblers kept my binoculars busy, and soon, sitting on a mangrove branch, was a dark bird of prey—smaller than a red-tailed hawk. Could it be the target species, a short-tailed hawk? Before I could get a better look, I heard a dull *thunk*, *thud*, followed by a sharp hiss. I jumped back and saw an elliptical dark mass on the leaf-littered ground. The "mass" and I soon recovered our wits, and it turned out to be a juvenile raccoon that had been pushed out of the tree by its sibling. Both raccoons were watching us. Perhaps even more surprised than the raccoon was an opossum partially hidden by the leaf litter. The hiss was triggered from the opossum by a startlingly close miss by the tumbling raccoon! It may even have grazed it. The raccoon scrambled back up the tree, no worse for the experience, to continue jostling with its sibling. The opossum stealthily ambled away, one paw in front of the other, its tail sticking straight out. Of course, the mysterious bird of prey had disappeared during all the commotion. I looked at Michael (he, too, did not have a camera) and shook my head. "Did we just see that?" Some things just cannot be predicted, and even when you experience them, you are left slack-jawed in disbelief. I noted that this was more reason to never hike alone—you may need a partner to back up those most curious encounters in nature. Otherwise, no one will believe you.

THE SCREAM

MICHAEL JEFFORDS

An author friend of mine, Stephen Lyons, wrote in his book *A View from the Inland Northwest*, "Rare is the time when we can quiet our inquisitive minds sufficiently and enjoy the present tense. Rarer still is knowing which quick hours in a long life will be the kind of precious touchstones we will draw on in later life." This elegant quote helped me remember a moment when I was exploring part of a lowland forest of giant bald cypress in southern Illinois. The area was littered with ancient trees between a thousand and fifteen hundred years old. Several were more than thirty feet around at their buttressed bases. Each tree also had a myriad of knees, structures that grow upward from the roots and are thought to help anchor the tree in the swampy soil, analogous to the flying buttresses of Gothic churches. The distinctive pattern on one ancient and massive knee grabbed my attention. The design in the flowing red wood was reminiscent of an iconic painting by Norwegian expressionist Edvard Munch—*The Scream*. Created in four versions between 1893 and 1910, these paintings are some of the most sought-after and highest-priced artworks on earth. These images have been the target of art thieves over the past few decades. Fortunately, all the paintings have been recovered.

The image of Edvard Munch's *The Scream* (*inset*), embedded in my mind as a child, sprang into my consciousness when I encountered this likeness, rising from the muck and mire of a southern Illinois swamp.

My first encounter with the "art world" was as a child through a most unlikely means—the local A&P supermarket in Paducah, Kentucky. It was my good fortune that my mother was addicted to acquiring books from this unlikely source. Growing up in rural southern Illinois provided little opportunity to explore world culture, except through these books. Each Saturday, during her excursion for groceries, she would acquire the "latest" intellectual installment available from A&P—these included a complete set of *Funk & Wagnalls* encyclopedias and a massive tome entitled *Worldwide Art*. The art book was sold in one-hundred-page increments, starting with Stone Age cave paintings and proceeding through seemingly endless supplements, each for the same bargain price of ninety-ty-nine cents—the same price as the encyclopedia volumes. We managed to acquire both sets, including the three-ring binder to organize the worldwide art collection. This was at the beginning of the television era—in the neighborhood, only my aunt possessed a television—so most of my evenings were spent poring over both the encyclopedias and the full-color art book. Like most massive collections of art, mine was replete with limitless examples of religious imagery from the Middle Ages, with mind-numbing variations of Madonna and child. I eagerly jumped ahead and was startled by the painting *The Scream*. To a seven-year-old, it was terrifying, disquieting, weird, but very exciting to see.

Later in life, years removed from my childhood pastime and well into my scientific career, I obviously had not forgotten that arresting image. As I traversed the eerie, otherworldly expanse of swamp, happening upon this unique cypress knee, I remembered my mother's efforts to broaden my horizon and was grateful to both her and the local A&P for expanding my world.

COLORFUL KATYDIDS

MICHAEL JEFFORDS

We all know katydids are green so they can blend into foliage. Well, that is only mostly true, as illustrated by the pair shown here. The phenomenon explaining the exception to the rule is called "erythrism" and is defined as "an unusual reddish pigmentation of an animal's fur, hair, skin, feathers, or eggshells." Pink katydids were first noted in 1874, with

an erroneous explanation that they change color seasonally to hide in fall leaves. That explanation was soon replaced by a suggestion from William Morton Wheeler that the color form was determined by genetics, similar to albinism. But how does that explain a bright-yellow katydid with green eyes? To understand, we must visit the realm of Mendelian genetics and journey to the Audubon Butterfly Garden and Insectarium in New Orleans. The insectarium displays scores of exotic living insects and features a research staff interested in many aspects of entomology. Through a series of breeding experiments, they postulated that the bright-pink color of katydids may be the result of a mutation and is heritable. Their experiments were reminiscent of Gregor Mendel and his peas and demonstrated that the even rarer colors yellow and orange must also be added to the definition of erythrism. In genetics we often associate mutations with the least adaptive and hence uncommon traits in animals and plants. In this case, that assumption appears to be correct, as these insects stand out like the proverbial "sore thumb."

This adult yellow morph of a katydid was found in a forest habitat (*top*). The pink nymphal katydid occurred in a late-spring prairie (*bottom*).

Katydids occupy a relatively low level in the food web. Either they feed on green plants, or some species are omnivores and eat a varied diet. However, all are important and easy prey items, if discovered, for larger animals, especially nesting birds. A katydid's only potent defense against a larger predator is crypsis. Because most live in heavily vegetated forests or grasslands, green is certainly the most appropriate color to blend into their surroundings. It seems likely that individuals bearing colors other than green do not survive long in the population, as they are extremely conspicuous.

Over the years, I have traveled widely and photographed many thousands of insects, but on only a handful of occasions have I encountered these uniquely colored katydid morphs. I found the pink nymph with the green eyes one year in a southern Illinois prairie as it hopped through the tall vegetation. I noticed the nymph because it was highly visible and probably not long for this world. A second occurrence of "katydid erythrism" occurred in a dark, forested prairie grove in central Illinois. The bright-yellow adult stood out like a beacon in the forest; I had no idea how it managed to survive to adulthood. A glance at the continuing research on katydid color variation from the Audubon Butterfly Garden and Insectarium, however, yielded this interesting statement: "The body color changed from green to yellow or orange over successive molts, resulting in a yellow or orange coloration as an adult." Perhaps this bright-yellow adult survived by being a green nymph through most of its life cycle. My guess is that after I immortalized it in my photo, it soon disappeared into the crop of a hungry bird!

SCAVENGER HUNT

SUSAN POST

During everyday life, and when traveling, we often witness the dog-eat-dog world of the food chain. Almost daily in our backyard, a Cooper's hawk perches and waits for an avian misstep to provide an easy meal. In Africa we witnessed a female lion kill a springbok and a lengthy standoff between a lion pride and a lone bleeding Cape buffalo that eventually escaped. Such encounters between predators and prey are both exciting and disturbing to watch. But what about those often ignored and much-maligned other members of the food chain—the decomposers? Who are they?

Vultures scramble for position on a Cape buffalo carcass in South Africa (*top*). Queensland Museum scientists scavenge a humpback whale (*bottom left*). A carrion beetle works on a skunk carcass in Wisconsin (*bottom right*)

A highlight of a South African trip was an encounter with vultures at a Cape buffalo lion kill. The lions and hyenas were gone when we arrived, leaving a gristly mass of buffalo muscle, sinew, and bones—its teeth locked in a macabre grin. As I watched, I could hear the vultures rip, pull, tear, and hiss—the carcass jerked with their efforts. Vulture hierarchy was on full display, with white-backed vultures seeking soft flesh and intestines, while hooded vultures tore at tendons and muscle inside the carcass. Dominant birds were in total possession, probing and feeding on the "good parts." Off to the side skulked satiated birds, while impatient, hungry interlopers constantly prompted defensive wing flapping and flying about by the dominants. By the end of two days, the buffalo was nothing but scattered bones and a grinning skull.

In Illinois our vultures are more skittish, and we seldom see them except as large "kettles" high in the sky. Here a different scavenging hierarchy is more easily viewed—with insects as the dominant scavengers. Flip over a roadkill raccoon or opossum and look beyond the maggots (fly larvae) for hairy rove beetles and carrion beetles, the adults of which feed on the ever-present maggots. A roadkill skunk in Wisconsin displayed several species of beetles navigating in and out of its nostrils and eye cavities—scavenging and cleaning the carcass. With no danger of lions or hyenas, it was easy to venture in for a closer look at this smelly diminutive world.

Now consider decomposers of a different ilk—humans. In Australia we spent a day on Stradbroke Island, off the coast of Brisbane. We were there to hike, look for birds, and see the spectacular flowers of *Banksia*. During lunch at a pizza parlor, patrons discovered we were Americans and hastened to ask us, "Have you been to the beach, to see the whale?" A humpback whale had beached and died on shore and was being dismantled by museum scientists. The pair of local women led us to the site. We saw dump trucks loaded with a brown Jell-O-like substance, a crew dressed in rubber boots and garbage bags, and a giant, fragrant whale carcass. We watched a crew from the Queensland Museum scavenge the whale. This was a unique opportunity for the museum to acquire a complete skeleton. They flensed the carcass (removed the outer blubber layer) with sharp knives, and the blubber was loaded with a rented end-loader into dump trucks (the brown Jell-O) and taken who knows where.

After viewing this unlikely drama, we returned to looking for birds and plants. I later learned that the museum was able to salvage the skull, skeleton, and baleen sheets. The museum's approach offered a different kind of decomposition, but one that proved just as effective as vultures and beetles at recycling the dead for the benefit of the living.

WORM CIRCLES

MICHAEL JEFFORDS

I had just arrived at work one morning when my phone rang, and a breathless colleague, Dr. Ed Zaborski, admonished me to "get over to the quad, pronto. You've got to see this!" He promptly hung up without telling me why. I was intrigued. Ed is a soil ecologist with a special interest in earthworms and usually not very excitable. I grabbed a camera and headed to the University of Illinois Quadrangle, a grass and paved commons surrounded by iconic Georgian buildings in the heart of campus. I arrived and noticed nothing out of the ordinary, until I saw a small area of sidewalk roped off with yellow construction tape. I approached the tape and... wow! Something peculiar was indeed happening. Maybe not "knock your socks off" front-page-news peculiar, but it was very odd. On the sidewalk were hundreds, perhaps even thousands, of large earthworms that had formed an "eye-of-the-hurricane"-type circle. Unfortunately, it was near midmorning, and most of them had dried up and died, but the scene was still *Twilight Zone* eerie. What was going on? Ed soon appeared and we speculated on possible mechanisms behind this phenomenon, but we were stymied, as there appeared to be no logical explanation. Our best guess was some sort of underground magnetic or electrical field had caused the worms to circle.

I had nearly forgotten about worm circles on the quad when, years later, I ran across the photos while scanning slides for a new project. Ed had since moved on, but I thought that maybe he had written up this

Stinkworms
(*Amynthas hupeiensis*)
surface on a sidewalk on
the University of Illinois
quad and form mysterious,
unexplained circles.

event in a scientific journal. A search of *Zaborski* and *worm circles* by our library staff yielded a typically obtuse but promising citation:

> Zaborski, E. R., and L. A. Soeken Gittenger. "*Amynthas hupeiensis* (Michaelsen, 1895) (Oligochaeta: Megascolecidae) in Illinois, USA, with Observations on Worm Circling." *Megadrilogica* 8, no. 4 (2001): 13–16.

Hmmm... It appeared Ed had figured out enough about worm circling to write a paper, so what had actually happened? First, *Amynthas hupeiensis* is an East Asian earthworm, thought to have entered the United States in the soil of Japanese flowering cherry trees that grace the Washington, D.C., Mall. The species had never been reported in Illinois, until shortly after sunrise, April 21, 1999—Worm Circle Day on the UI Quad. Early that morning, Ed had noticed a peculiar odor of "freshly dug carrots" from the worms that I had obviously missed. This odor has given the species the common name "stinkworm." This was great information, but what about the circles? While it seems that the quad had seen thousands of worms surface after a rainfall the night before, only in this one location were they circling. I was getting excited, as the paper seemed to be closing in on an explanation, until this one telling passage: "We have no suggestions, however, about what caused this one particular aggregation of worms to circle on April 21st." What? Like good scientists, Zaborski and colleagues measured the diameter of the circle, recorded the number of worms, and reported other corollary data. Great, but they provided no ideas on circles. Unfortunately, sometimes even a thorough study can't explain some of the things we see. The mystery remains!

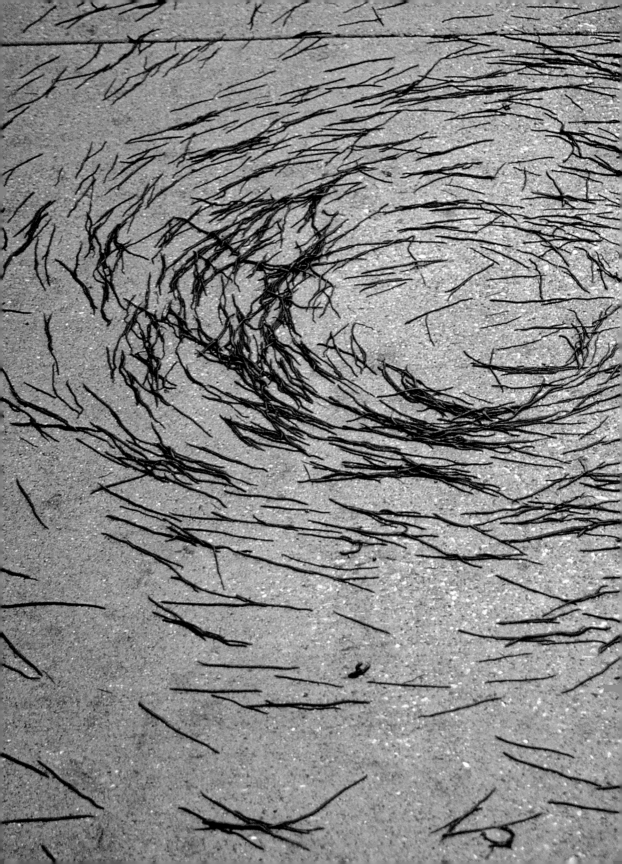

SIZE DOES MATTER

MICHAEL JEFFORDS

After spending almost five years in graduate school, working on mimicry in butterflies, my first entomological position was quite a shock—a soybean entomologist. "Eww," I thought then, but soon found the fieldwork to be relatively interesting. There is nothing more "rewarding" than taking endless sweep samples in the corduroy soybean fields that blanket much of central Illinois. Upon occasion, though, fieldwork was downright bizarre. In the early 1980s, before the advent of herbicide-tolerant soybean varieties, weed control was a major issue, and most growers either hired "bean walkers" or coerced family members to traverse the endless miles of soybean rows with weed hooks, hoes, or bare hands to remove all the unwanted vegetation. Velvetleaf, giant ragweed, cocklebur, lamb's-quarters, Canada thistle, and, of course, common milkweed were all targets of the razor-sharp weed hooks. I was definitely more interested in bean-leaf beetles, Mexican bean beetles, green stink bugs, and velvetleaf caterpillars, but the abundance of weeds was hard to ignore.

This photo is a re-creation of what I saw that day. Thanks to the miracle of Photoshop, you can experience an abundance of *Labidomera clivicollis*. As it happened, I had no camera with me to record this remarkable observation.

One of the most unusual events in my five-year stint as a soybean specialist began early one morning with a phone call from one of our growers in the East St. Louis region. He began the conversation with a chuckle and an invitation to "hop on down to Collinsville," as he had something to show us. "You won't believe it" was his parting remark. My colleague and I were headed to the area the next day anyway, so off we went, our curiosity piqued by this cryptic summons. Our relationship with growers was always cordial, yet this individual was a favorite and we had no idea what to expect. Upon arrival we found that all of his soybean fields were harvested and the grain had been taken to the local elevator for sale. A single grain truck stood like a lonely bull in his farm yard, its load covered by a heavy tarp. Bob, the grower, met us with a grin, a bone-crushing handshake, and the admonition to jump up on the sides of the truck and help him remove the tarp. "What's up?" we asked. "This load of beans, while some of the cleanest I've ever harvested, was rejected at the elevator yesterday, and you won't believe why!" With a flip of the tarp, we were both pretty much speechless by what was in the truck. The surface layer of the soybeans was alive with bright-red and black milkweed beetles (*Labidomera clivicollis*), thousands of them, crawling back and forth across the yellow beans. No wonder this truckload was rejected: the beetles feed on milkweed and have cardiac glycosides in their bodies that deter predators—not so good for a food crop!

So what's the story here? The field that these beans came from had a late flush of common milkweed that Bob had ignored, knowing they would pose no problem for his modern combine. However, he did not realize that this field would become a magnet for milkweed insects, and these beetles had grown to enormous numbers in his "unofficial milkweed preserve." Being the same size as a soybean, they were harvested along with the crop and had crawled to the top of the load just in time to be seen by inspecting personnel. Indeed, this was worth the trip, and truly bizarre.

SIXTY-MINUTE CRUISE

SUSAN POST

Asian carp are introduced fish. They were brought to the United States to clear algae and parasites from catfish ponds in Arkansas and Mississippi, but floods washed the carp into the Mississippi River. They soon worked their way upstream and now thrive in the Illinois River. Asian carp consume up to 40 percent of their body weight in plankton each day and outcompete native fish species.

Some events are so bizarre that only a cartoon can depict the experience (drawn by Joseph Spencer)! Jumping Asian carp (photo by Levi Solomon).

During the summer of 2007, I organized a field class focused on sand prairies and their butterflies and skippers. The class was located along the Illinois River, so for evening edification I contacted the Forbes Biological Station and director Dr. Greg Sass to arrange a boat tour so the class could "experience" Asian carp. Our evening cruise was to feature the ability of carp to leap out of the water when startled. I had seen videos of fish leaping around boats, so I thought people in a field class on butterflies should also be exposed to "flying" fish. After dinner Greg announced our intentions. The trip was optional, but the response was overwhelming. Everybody desired to go, so more than one cruise was needed. I thought to myself, "If I play my cards right, I might not have to participate." While I wanted the class to see this phenomenon, I wasn't that keen to go. Unfortunately, there was room for me on the final cruise. Reluctantly, I climbed aboard and found myself a seat in the left rear of the boat. I thought, "Great, I'm in the splash zone!" My friend Karen was seated next to me. We slipped into the calm water with the setting sun. Periodically, I heard thumping—carp hitting the

bottom of the boat—but otherwise the scene was idyllic, and I was happy to be on the cruise. Greg told the group about the carp and his recent experience with a CNN news crew. As soon as the fish started to jump, a female reporter became flustered and started to scream. Karen then smirked to Greg, "I bet she had on high heels and lipstick!" Greg smiled and shook his head.

We met a second boat in the middle of the river, one outfitted with electroshocking equipment. At Greg's signal, his colleagues started the generator and put the electrodes in the water. More than fifty fish leaped from the water in a seemingly choreographed, but very bizarre, synchronized swimming routine. Then smack, they hit the water, but kept jumping as our boat moved along. What a sight! Greg carefully piloted, attempting to slip between the jumping fish. Suddenly, I was hit from behind by a large carp that scraped across my face, knocked my glasses off, whacked my right thigh, and landed, bloody and wriggling, in the bottom of the boat. Like the fish, I was stunned. Karen, however, started to scream, flailing her arms and legs. Why, I don't know—I was the fish's victim! I, on the other hand, just stared at the white slime on my pant leg, still trying to comprehend that I had been smacked in the head by a ten-pound fish. My glasses had landed beneath Karen's flailing feet and became somewhat bent. After reassuring Greg that I was fine, we headed back to shore. Karen, like the leaping fish, eventually ran out of energy and calmed down. I still remember the look Greg had given Karen when she made fun of the screaming reporter. It was a look that said, "We'll see." I am proud that my reaction was one of calm dumbfoundedness, and now I can add "getting pummeled by a ten-pound carp" to my life's list of accomplishments!

LIFE LIST—YES OR NO?

SUSAN POST

I started birding in the 1990s, and I keep lists—a life list, an individual outings list, and a yard list. While my yard list is from the middle of the "exotic" town of Champaign, Illinois, it is far from uninteresting. As my avian self-education efforts ballooned into an obsession, I accumulated field guides, top-of-the-line binoculars, and a spotting scope, and I now frequently plot truly exotic trips to see new birds. Michael, while amused, did not attempt to squelch my newfound passion. We could travel anywhere, as long as there was something interesting to photograph.

The elusive varied thrush, a bird of deep conifer forests of the Pacific Northwest, made a surprise appearance in Champaign, Illinois (photo by Patty Dickerson). Its tenure here, unfortunately, was short-lived.

In 2001 we visited the Canadian Rockies with a goal of seeing glaciers and wildflowers. Our final stop was Mount Revelstoke in British Columbia—a park of lush rain forests and mountain peaks. While setting up camp, I heard a loud whistle, similar to that of an English bobby, but saw nothing. With camp set up, Michael and I headed out to explore. Each trailhead sported an interpretive panel, seemingly ripped from a naturalist's journal. One illustrated a varied thrush and described its whistle-like call. Could this be the bird I had heard? The taunting whistles continued, but my diligent searching yielded nothing. A varied thrush resembles, as one birder wrote, "a robin that has been decorated for Halloween." They are birds of the Pacific Northwest, favoring dense, moist coniferous forests. I mentally marked this as a species I needed to see. My next opportunity came when Michael and I were headed on an Alaskan cruise. During a day in Vancouver, I searched an ancient conifer forest with no luck. The bird had now reached nemesis status.

On the evening of October 23, 2014, Michael and I were in Springfield at the Illinois State Museum, speaking about encounters with rare and unusual birds, for the opening of an exhibit of original Audubon bird prints. When we appeared at work the next day, our neighbor and colleague Patty Dickerson questioned, "Where were you last night? I tried to call you." She then pulled out a photo. "Look what was in our garage." Stunned, I was looking at a varied thrush! Patty explained that while sweeping out her new garage, she heard a commotion in our yard, followed by a *thunk*. She opened her garage door and saw a gray robin-size bird fly inside. Outside the garage door, a frustrated neighborhood Cooper's hawk flew off. When the bird inside the garage flew into the rafters, Patty noticed that it was not a robin and snapped a picture. Her attempted call to me with a tentative identification yielded nothing. Her husband, not wanting bird poo on his new garage floor, banished the bird from the garage. After seeing the photo of the correctly identified varied thrush, I dashed home, staked out our backyard, played the plaintive-whistle call, and scrutinized every robin. Nothing. Later, by our brush pile, I noticed an orange and gray pile of feathers. I still have not seen a living varied thrush, which leads to my dilemma. Could I add a "yard bird" to my life list, the remnants of which I had clutched in my hands?

JUST BE CURIOUS

In *East of Eden*, John Steinbeck wrote, "I take a pleasure in inquiring into things. I've never been content to pass a stone without looking under it." In that one simple statement, Steinbeck summed up the life of a naturalist-biologist. This is what we are trained to do—turn over the rocks that we encounter to see what wonders lie beneath. We analyze and try to understand why things are the way they are. We ponder things that we have not yet encountered and go out of our way to see and experience unfamiliar things. The field biologist's mantra is a simple one: "The worst day in the field is still better than the best day in the office!" This love for the novel and the wild animates the lives of field biologists. Is it any wonder that we greet each day with curiosity? How will a full solar eclipse affect local bird behaviors? Why are there so many species of antelope in Africa? Why are penguins so curious? What do blackish oystercatchers feed their chicks? Why are some butterflies so common and others fleetingly rare? The list of curiosities and questions is infinite, and a fresh encounter in nature is all that is needed to ignite the fires of observation. Without curiosity, why would we have journeyed to remote Brazil to see the world's largest rodent? Or stopped on our honeymoon to explore Grandma's bog? Or driven into the bleak unknown of the Namibian wilderness to find a "tree" with only two needles? Our curiosity is fueled by a desire to see, to know, and to depict, and it has always been that way. This series of essays has no common theme, other than the fact that we were curious enough to stop and explore. What most others would have passed by—those tied to a daily "to-do" list—we view as an adventure, a challenge, or simply a reason to "just be curious."

In the Falkland Islands, we observed blackish oystercatchers feeding a shellfish to their chick—mystery solved (previous page). Curious rockhopper penguins investigate a GoPro camera situated on the edge of their colony in the Falkland Islands (left).

BEE-EATERS!

MICHAEL JEFFORDS

Crossing the border from northern Namibia into Botswana was surprisingly uneventful—a dusty, nondescript roadside building resembling a small-town post office beckoned us to enter. We showed a friendly lady our passports and filled out yet another meaningless form regarding our intentions in this new country. With a resounding visa stamp of approval, we were off to explore one of the world's most famous sites, the Okavango Delta! Actually, we were going only as far as the horn of the delta, but that was far enough, as we were at the end of a very long birding trip in this remote part of southwestern Africa. We had visited such signature locations as the great dunes along the Skeleton Coast, sea-kayaked with endangered Cape fur seals, and tent camped among wild throngs of chacma baboons. It had been quite an adventure.

White-fronted bee-eaters decorate a dead tree limb like so many Christmas ornaments. As it was near dark, a powerful flash enabled me to capture their incredible colors.

Our final destination was the Shakawe Fishing Camp, located hard on the shores of the Okavango River, a clear, pristine stream that arises in Angola and flows for nearly a thousand miles across northern Namibia and into Botswana, only to disappear into the vastness of its delta in the Kalahari Desert. The river has no outlet to the sea. Earlier in the trip, we had taken boat rides on the river where it formed the border between Namibia and Angola and had experienced close encounters with hippos, Nile crocodiles, and innumerable species of aquatic birds that call the papyrus-lined paradise home. As we passed a steep cut in the riverbank, laced with hundreds of holes, our guide said this was a breeding colony of the stunning carmine bee-eater, a migrant that would not arrive until August. This was a disappointment, as we, unfortunately, were there in June.

As we continued our road journey, we passed dramatic Popa Falls, where the river plunges twelve feet over a series of river-wide boulders, stopped for an afternoon at a private game reserve, and arrived at our destination late in the afternoon. Our lodging was on the river, and we were serenaded all night by the deep, sonorous calls of hippos. The reason for our visit was an attempt to find the legendary Pel's fishing owl, the only owl that feeds exclusively on fish. This giant, described as "resembling a cocker spaniel sitting in a tree," does not need to silence its wings like other owls, because it plucks fish from the river. Luckily, the next morning we found our quarry, sitting high in a tree, looking curiously at us.

Later that day, as we hiked through the riverside thickets, seeking other avian quarry, I bored of the quest and wandered off from the group. I headed downstream, lazing away the afternoon, when I noticed along the distant riverbank quite a flurry of activity. It was almost too dark to see clearly, so I headed closer. The sun was just fading below the horizon when I arrived at the riverbank to find hundreds of bee-eaters, coming and going in a chaos of color! They were not carmines, but white-fronted bee-eaters. It was nearly dark, but fortunately, I had a powerful flash on my camera, and I took chances with the exposures. I took many shots as the birds came and went. Imagine my delight when the images showed off the birds' dramatic colors, like so many Christmas ornaments decorating a tree!

CAPYBARA ENCOUNTERS

SUSAN POST

Since we have been married, Michael and I have had a succession of neotropical companions—guinea pigs. Our first one appeared as a Christmas gift, and I instantly bonded with this small, quirky rodent. My love affair with these animals has also led to an affection for all rodents, especially the larger examples. I keep a squirrel checklist, just as birders keep life bird lists. Imagine my surprise when a scientist friend regaled me with stories of a large pig-size rodent that resembles a guinea pig that he had seen while studying in South America. He had observed capybaras, a rodent John Kricher describes in *A Neotropical Companion* as "a magnificent creature to behold." Seeing a capybara in the wild immediately was on my list of things I had to accomplish. A Peru birding journey yielded a distant sky-eye view from a canopy tower of a capybara family—a trio wallowing and ear flicking—and it proved to be totally unsatisfying. It was not until we journeyed to the Brazilian Pantanal that I was able to satisfy my curiosity. While bumping along the rutted Trans-Pantanal "Highway," our first stop was a drying wetland. There, mingling with a bevy of great white egrets, was a capybara! Déjà vu kicked in, as I had seen this scene before, only in Africa, and the players were hippos and egrets, not giant rodents.

A capybara family group heads out from the nearby river to spend the morning grazing on the Brazilian Pantanal. Note the blooming pink piuva trees in the background (*top*). A very lucky capybara with a healing wound from a jaguar plays host to a black-capped donacobius (*bottom*). A capybara track is embedded in the riverbank (*inset*).

The next morning I awoke and looked out my window. Imagine my surprise to see a small group of these enormous guinea pig–like creatures grazing and lounging in the nearby field. Pink piuva trees and a tropical river formed a stunning background. Quickly dressing, I stealthily walked as close as possible. I startled one, and it gave a wheezy grunt as it hurled itself into the river, landing with a large splash. The rest seemed unfazed by my presence. The capybaras looked and smelled earthy, and in my mind they were equivalent to rotund hippos, only in South America.

Capybaras are stocky, with short, thick legs, and they are always found near water. They are semi-aquatic and are found in and along rivers and lakes large enough to have open sky above, with a margin of aquatic plants or grassy vegetation for their grazing. To escape predators, they can swim underwater and hold their breath for long periods. Their genus name, *Hydrochoerus*, is Greek and translates to "water pig." On one of our later boat rides, we encountered a lone capybara on shore that had obviously experienced a close call with a jaguar. Both sides wore symmetrical jagged flesh wounds. The healing claw marks attracted the attention of a bird, the black-capped donacobius, and it was riding on the capybara's back like an African oxpecker, pecking at the wound.

Later, another capybara encounter reminded me of Africa. While I would not want to surprise a potentially dangerous hippo along an African river, under similar circumstances a capybara encounter in the Pantanal was delightful. As I watched a family group walk to the river, mom in the lead, dad among the offspring, I counted myself lucky. I had spent quality time with the world's largest rodent, my newest neotropical companion, and a truly "magnificent creature to behold."

WELWITSCHIA!

MICHAEL JEFFORDS WITH SUSAN POST

We entered a landscape that passed from dry savanna through mountainous passes to grassy expanses to, well, nothing. We were traversing central Namibia in search of birds on our first trip to Africa. Sue's journal entry reads, "We head out to the lichen fields (which resemble the moon or, closer to home, the South Dakota badlands) searching for Welwitschia, a horizontal plant that is low to the ground. We locate a fine specimen in the 'forest' that is delineated by rocks, but soon leave this 'moonscape' of nothingness. There are more larks to find!" Driving dozens of miles out of the way, over some of the roughest roads on the planet and through the most desolate, hostile landscape on Earth to see a plant, albeit endemic and world famous (at least to botanists), may seem the height of stupidity. Not surprisingly, we endured the grumbles of our less than enthusiastic traveling companions. Yet Sue and I insisted. As biologists, we could not come to remote Namibia and not see Welwitschia, the world's oddest tree. The "trees" occupied the most widely dispersed forest on Earth, as we saw only a single individual in our forest search.

An ancient *Welwitschia* tree grows on the vast Namibian desert plains. Its nearest companion in this highly dispersed "forest" was at least a mile away.

The *Welwitschia* tree is a conifer with only two dull green leaves that resemble leather, barbershop strops that emanate from a central dark-brown trunk. The leaves are shredded by the incessant desert winds. We can think of no more fitting plant to grace this alien landscape. *Welwitschia*, an ancient organism, is "monotypic" across its taxonomy. This means that it is one of a kind and is a true "living fossil." *Welwitschia* is closely related to other conifers—pines, spruces, larches, and various firs—and fossil evidence from South America indicates that it likely arose in moister conditions during the lower Cretaceous. The two leaves of *Welwitschia* grow continuously and may reach eight to twelve feet long. An old tree can "tower" up to five feet in the largest specimens; ours was less than a foot tall. While most conifers are wind pollinated, a male *Welwitschia* depends on insects to transfer its pollen to the nearest female. Given that a female may be some distance away, this is a remarkably adaptive strategy in the vastness of the Namibian desert. Individual plants are extremely long-lived, and estimates put most large examples at around a thousand years old, with extreme specimens approaching two thousand years in age!

Ironically, this was not our first sighting of this unique and marvelous plant, as some years before we had viewed a rather forlorn-looking specimen residing in a large pot in a domed greenhouse at the Missouri Botanical Garden in St. Louis. It turns out that seeds for *Welwitschia* can be purchased from specialty seed dealers. Seeing *Welwitschia* in a greenhouse was the equivalent of once viewing an Indian tiger "proudly" displayed in a cage at our local pet store. Both experiences were underwhelming. We were surprised to later find out that in this seemingly uninhabited and uninhabitable landscape, threats to *Welwitschia* existed and included overgrazing by zebras and rhinos!

THE STALKER

SUSAN POST

It was my first day on the Falkland Islands, and our group of five headed to Volunteer Point. I felt as though I was not supposed to know the point's location—and I didn't—as we piled in the Range Rover at one thirty in the morning. It was pitch-dark. We bumped and bounced along a dirt track for more than three hours, seeing nothing but sheep in the headlights, and listened to our driver recount her Falklands War experience. Volunteer Point is a privately owned farm where sheep intermingle with penguins. The area contains a two-mile-long beach, and Volunteer Point is home to the Falklands' largest king-penguin colony—around one thousand birds.

A younger juvenile with full woolly coat occupies the edge of the colony (*top*). My "stalker" chick bonded to me for the day (*bottom left*). An adult king penguin with full yellow-head markings preens its feathers (*bottom right*).

Prior to our foray to the Falkland Islands, one of the books I read was *The Moon by Whale Light*, by Diane Ackerman. In it she describes an encounter with a king penguin while wearing a yellow sweat suit. The penguin followed her about until it was convinced that she was not of its kind. King penguins have bright, graduated orange-yellow comma-shaped ear patches that extend down onto the breast. The penguin "saw" yellow and was likely hoping for a mate. After reading that, I quickly went to our local mall in search of a yellow scarf that I hoped would mimic penguin yellow.

We arrived at Volunteer Point at four thirty, and as we crested the final hill, tall forms—initially mistaken for massive numbers of cruise-ship "explorers"—materialized into king penguins. Their erect white bellies glowed like a full moon against the brown-green hills. Our driver left us and promised to return twelve hours later. We were in the midst of organized penguin chaos. Colonies of gentoo penguins graced the tops of circular rises they had created by hundreds of years of accumulated penguin debris. Magellanic penguins brayed in pairs and small groups, poking their heads from within deep, mysterious burrows. Several large colonies of king penguins graced the gentle hillsides, organized by age. The adults were in the center, surrounded by a moat of multiage chicks. The periodic trumpeting of adult king penguins sounded like tin New Year's Eve horns. Young juveniles stoically stood on the edge of the colony, mimicking humans in furry brown coats. Older chicks sported newly grown feathers with bodies still laced with tufts of brown hair—like hairy men at the gym.

The weather was a mix of rain, sleet, and wind. I wore several layers, including a dark raincoat and pants, ornamented with my yellow scarf. From a distance, I could pass for a king penguin. For the first two hours, I was dumbfounded by the scene: What should I do, and where do I go? What was proper penguin etiquette? Finally, I accumulated nerve enough and "flirted" with the king penguins by leaving a piece of my yellow scarf to blow about. I sat down, watched, waited, and listened. The most curious chicks soon came to investigate, and penguins surrounded me. They soon decided I was not one of them, lost interest, and continued about their business. All, that is, except for one. As I got up to leave, a shaggy juvenile followed me. As I explored the landscape, my stalker was always just a few feet away. I had my companion for the day.

LADY'S SLIPPERS FOR SALE!

SUSAN POST

The most remarkable experiences are often totally spontaneous. This was the case as we drove along a busy highway, heading north from Petoskey, Michigan, on our honeymoon. We had been shopping to restock our camping supplies and searching, as the tourism brochure stated, for beaches "littered with Petoskey stones." Yeah, right!

Arethusa orchids graced nearly every hummock in Grandma's bog in northern Michigan. Buckets of lady's slippers (*left*, drawing by Patty Dickerson) caught our attention as we drove by.

LADY'S SLIPPERS FOR SALE, the sign read; cans of pink, white, and yellow blooms were a blur as we whizzed by in our car. This deserved a second look. We stopped, and I spoke to the grandmotherly lady sitting in the lawn chair. The hundred or so blossoms in the old tin cans attested that this was indeed an unusual place! Yes, orchids were for sale—two dollars a dozen or five dollars for rootstock. Not interested in cut blooms, I envisioned large clumps growing in the wild and dared ask if we might see them. No, she was afraid we would get lost in her "swamp," a quarter mile "back of the house." I took her to meet Michael, and he convinced her that "swamping" was second nature to us.

The grandmother's fears allayed, we headed down the path, through the muck, to enter not a swamp but an ancient bog. No open water was to be seen, and tamarack trees grew seven to eight feet tall—the mat was spongy, wet beneath us, but firm. Before the flower harvests, there must have been massive clumps of the large showy lady's slipper blossoms, yet numerous single plants or twosomes remained, littering the sphagnum. Wood lily and yellow lady's slipper blossoms confirmed our earlier suspicions about the diversity of Grandma's bog. Most spectacular was the *Arethusa* orchid in hot pink, its startled expression on almost every sphagnum hummock.

Exploring done, how do we get out? After several wrong turns, holes of black water, and branches across the face, our very wet feet found the path, and we sloshed our way out. After we checked back in, the grandmother said she had been worried and was glad to see us return safely. She hadn't visited her "swamp" for five years (her grandkids had harvested the blooms), but was curious to know what we saw and did we see such-and-such. She thought a boardwalk would be nice, but it cost too much, and for a million dollars she'd sell the property. Like the old man of the bog (see "Mr. Jackman's Bog"), she appreciated her "swamp" and exploited only a small part.

A return trip to Michigan ten years later found us driving down the same, but even busier, highway—this time very slowly, looking for Grandma's house and her bog. We nearly missed the ramshackle, deserted house, now surrounded by various commercial and residential developments. Grandma obviously was no longer living there, but was her bog still intact? We were stopped from finding out by an imposing chain-link fence. Unfortunately, from the look of the area, the bog was likely gone. If we pass that way again, though, we will certainly keep our eyes open. Perhaps one day, a young entrepreneur will be stationed in front of the subdivision, not selling lemonade for fifty cents a glass, but with a big sign: LADY'S SLIPPERS FOR SALE!

Lady's Slippers
for Sale!

FLYING SQUIRREL CAFÉ

SUSAN POST

I keep a squirrel life list, much to the amusement of Michael and my birding colleagues. Michael endures stops at almost every prairie-dog town or chipmunk habitat, and I constantly lag behind on hikes to investigate squirrel middens and errant squeaks. During my first birding foray to Texas, while the group searched the Brownsville airport for the Tamaulipas crow, a rare Mexican visitor, I happily discovered a Mexican ground squirrel. The crow was a no-show, but I added a new squirrel to my list.

Sue admires the Flying Squirrel Bakery-Café sign, forgetting that she has a bag full of pizza and cookies in grizzly bear country!

During 2011 Michael and I were asked to lead a cruise for the University of Illinois's Osher Lifelong Learning Institute (OLLI) to Alaska. Never having been there did not deter us, and I conducted extensive research to find places and excursions in which the group might be interested. After cruising the Inside Passage, we headed to Denali, with a stop in Talkeetna for the evening. While in Talkeetna, everyone had free time. During my research, I had discovered a restaurant just three miles from downtown—the Flying Squirrel Café. It billed itself as a humble bakery-café at the northern reaches of flying-squirrel habitat—pastries and wood-fired pizzas in the semiwilds of an Alaskan forest and a chance of seeing flying squirrels! For a squirrel-list keeper, this destination was second only to Denali, which on most days is enshrouded by clouds.

From our lodge, we walked to downtown Talkeetna, a small, quirky village, seemingly frozen in time, that supported a plethora of hippie wannabes. A quick stroll past the few stores led us down to the river—Talkeetna means "where the rivers converge." The Flying Squirrel Café was in the opposite direction, so with no public transportation we walked the three miles on a paved bike path. The walk was enjoyable, the greenery lush, and the daylight lengthy. Upon our arrival at the café, I purchased the requisite T-shirt and ordered a wood-fired pizza and salad. Cookies were purchased for our bus trip the following day. As we prepared for the trek back, we overheard a group of locals discussing how Alaskan chickadees are smarter than Colorado ones. They saw my camera and binoculars and informed us we should go to XYZ Lakes to look for loons. These folks used to run a nature tour company; they thought we would enjoy the scenery of this area that was good for bird-watching.

Unable to finish our pizza, we packed the leftovers in a to-go box and headed out for the three-mile walk back to our lodge. At an intersection, we wandered into XYZ Lakes for a quick look. While admiring the first lake, I noticed we were in a wilderness area, and posted near the entrance was a sign that a bear had been sighted during the past July 4 weekend. We went no further, as we were carrying a partially eaten pizza and a box of cookies in bear country! After a nervous walk back, we emerged, unscathed, at the lodge by ten. As it was still light, we enjoyed almost cloudless views of Denali.

Two days later, a couple of our OLLI members shared an article from the local newspaper. A grizzly sow with a cub had attacked a group of teens participating in a National Leadership School course not far from where we had hiked with our pizza. Where ignorance is bliss... and I failed to see a single flying squirrel.

TO SEE A GIANT OTTER

SUSAN POST

If wetlands are the heart of the Pantanal in South America, then rivers are its aquatic arteries. For these past two days, we have skimmed the waters of the Cuiaba River, racing up and down its main channels and ever-narrowing tributaries. Sometimes the scenery is only a blur of greens and browns, with quick stops to photograph a shoreline—pink Piuva trees, hyacinth macaws, or Jabiru storks. Other times we must ease along, teasing out the area's treasures—Jacaré caiman, jaguar, or giant otter. The giant otter, like the jaguar, is a blend of the elusive and the showy. Photographing jaguars quietly stalking the densely vegetated banks was certainly a highlight, but by the end of the first afternoon, I had more jaguar photos than I could process. Seeing three in a day was both spectacular and unusual.

A giant otter munches on a large fish in the Cuiaba River that flows through the Brazilian Pantanal.

The next morning, prior to boarding the boat for a full day on the river, I have my Pantanal mammal card out and point to the picture of a giant otter. A smile to Ronaldo, our guide and boatman, is all it takes to solidify our quest for the day. He nods and smiles back. Later, after seeing yet another jaguar, he radios to find out where we might encounter an otter family. We soon happen upon a group among the water hyacinths, cavorting and fishing. Once they catch a fish, however, they enter a tangle to eat, disappearing from view. I hear them crunching, growling, and splashing.

We move on and soon find another otter family calling to each other. The sound is a combination of snarls, growls, and meows, like pinching a latex balloon. The radio crackles with a wild peccary sighting. We give chase, but are soon distracted by yet another otter family—this one close and oblivious to our presence. For more than thirty minutes, we watch as they bob, tussle, yelp, fish, feed, and growl—rambunctious otters in the harsh, bright sun. Shoreline branches interfere with composition, and trying to photograph from a rocking boat only magnifies the challenge. Ronaldo, with an adept maneuver, gets the boat so we have better angles for photos. He is so impressed with the otter performance that he whips out his phone for a couple of photos as well. I am close enough to study the unique patterns of cream-colored splotches on their throats. As they tussle with their slick, fishy prey, their webbed feet deftly bring fish to mouth. I am also aware of their size; the otters in this family group are huge—at least five feet long, and that's not counting the tail. It is not rounded like our familiar river otter, but flattens out like a beaver. Their genus name, *Pteronura*, is Greek and translates to "wingtail." Their Spanish name is *Lobo del Rio*—river wolf—as they are the feared pack hunters of these tropical waterways. Attacking as a group, they can even drive away a jaguar. One look into their eyes, and I see not the cute face of a river otter but one of a pure predator, almost evil.

The following morning brings one last river jaunt, and as a gathering of giant otters passes by, I glance and share a smile with Ronaldo. It's full throttle ahead.

WHO KNEW?

anguages constantly evolve as new terms and words are added to the lexicon. The phrase "Who knew?" is a relatively recent urban expression that is often used in a sarcastic manner to denote something that is considered to be obvious. In colloquial usage, usually on the Internet, this is appropriate, but when combined with scientific or natural history observations, perhaps the meaning should be broadened.

Like languages, science is a rapidly changing and evolving entity that depends on new discoveries and observations. In this chapter, titled "Who Knew?," we feature the term not sarcastically, but as a way to point out that sometimes commonly held knowledge and concepts often interact with nature in unexpected ways. The events range from the merely interesting to the unfathomable. Anyone in the right place and at the correct time can experience these events, but their significance and interpretation may go unnoticed. Oftentimes, repeat visits and familiarity with the players and underlying scientific principles associated with these unique or unusual occurrences are necessary before they provoke a proper level of interest. Who knew that horses could cause the demise of tiny crane flies, that simple streetlights could disrupt a farmer's harvest schedule, or that hotel swimming pools in the tropics would be an irresistible lure for visiting scientists who never so much as placed a toe in the water? Who knew that a large goat and an endangered species of otter would make acceptable house pets? The explanations and the scientific principles behind the essays presented here delve into genetics, plant physiology, biogeography, landscape and plant ecology, and medical pharmacology, and they even touch upon human sociology. As W. Mark Richardson so aptly stated in a 1992 article in *Scientific American*, "A Skeptic's View of Wonder," "As the island of knowledge grows, the surface that makes contact with mystery expands." Who knew that the earth could be so complex and wonderful a place and a source of never-ending discovery?

ALL THINGS CONNECTED

MICHAEL JEFFORDS

To quote John Muir, "When we try to pick out anything by itself, we find it hitched to everything else in the Universe." As I biologist, I know this to be true, but it is not so easy to witness and actually document specific cases. For someone who has been photographing Illinois's organisms and landscapes for more than forty years, visiting sites repeatedly makes this connectedness more apparent. One of my favorite areas in the Shawnee Hills of southern Illinois to photograph was Hayes Creek Canyon. Many years ago, this isolated canyon was the best spot to see and photograph dripways (channels made by water running across and down sandstone ledges). The ledges and channels supported an interesting set of wildlife, including a madiculous (rock-loving) crane fly (*Dactylolabis montana*). On quiet, cloudy days in early spring, usually after a rainy night, aggregations of these diminutive crane flies would gather on the rocky shelves. The males would clump together in groups (called leks), sometimes forming bundles as large as softballs, and roll around the rocky shelves to attract females. Any female crane fly that happened along would literally dive into the amassed males; the ball would undergo a frenzy of activity for several seconds, and then out would come the female with her chosen

Sandstone dripways (*left*) are a common feature of the Shawnee National Forest. Crane flies pair up on a dripway (*top right*). Dripways now have spring wildflowers replacing the brown algae (*lower right*).

mate. The male would encase the female in his long, spindly legs, fight off the attention of other males who were trying to intrude, and slowly "walk" his spouse over to a nearby dripway. He would then mate with her, and while imprisoned by his legs, she would lay eggs on the moist algae growing on the dripway. After she finished, both would fly off, ending the miniature spectacle. When the eggs hatched, the tiny fly larvae would wiggle their way up the dripways and spend their lives as predators in the leaf litter of the forest bordering the sandstone ledges.

I observed this phenomenon for many years and even published a scientific paper on it. A visit a few years ago revealed that everything had changed in an unfortunate way. A large horse camp was established nearby and created many new trails upslope from the sandstone ledges. During heavy spring rains, soil that had eroded from the trails now filled each of the dripways. They no longer supported brown algae. It was replaced with sinuous skeins of local wildflowers. The whole scene was beautiful and might seem natural to a first-time visitor, but I knew better. The crane flies were conspicuously absent from the "new" landscape, likely something only an entomologist would notice. They had ceased to lek at the site because their crucial dripways no longer provided suitable habitat for their existence—the landscape was altered by horses and human activity. Perhaps no one cares but me, but I see it as a diminution of nature, a connection forever broken. Hayes Creek Canyon has been eternally changed by the absence of the spring ritual of *Dactylolabis montana*.

THE GINGKO EFFECT

MICHAEL JEFFORDS

Most of us are familiar with the gingko tree because of its distinctively shaped leaves. Also, the aromatic and unpleasant fruit produced by female trees in late summer is difficult to overlook! These pink odiferous morsels often litter the neighborhood sidewalks and come home with us, imbedded in the treads of our shoes. Likewise, we have all heard of the unrelated phenomenon "light pollution" from magazine articles or science programs that bemoan the fact that many of us can no longer see the stars, planets, and galaxies in the night sky. How could these two very disparate topics be related?

A perfect ginkgo leaf (*above*). The Ginkgo Effect was created by streetlights altering the photoperiodic response of the nearby soybean plants. Note the direction of the lights and the effect of the shadows from the light poles (*bottom*).

Several years ago, while conducting low-level aerial photography over the agricultural landscape of central Illinois, I witnessed an unusual occurrence. But first, additional background is necessary to understand the observation.

The ginkgo tree, while native to China, has been widely cultivated and imported around the world (and into the United States as a street tree). It has virtually no enemies to feed on its characteristic leaves. The scientific name, *Ginkgo biloba*, refers to its bilobed, fan-shaped leaves. Amazingly, every leaf on a gingko tree remains perfectly pristine throughout its life, as no insects can feed on it. To continue our investigation, light, in the form of photoperiod or photoperiodism, affects us all, but we seldom appreciate its role in the world. A simple definition of photoperiodism is "the physiological reaction of plants and animals to the length of day or night." Among humans, some individuals respond to the shortened day length in winter with seasonal affective disorder. Victims experience major episodes of depression, beginning in late fall and often not ending until spring. Other animals react to changes in day length by going into hibernation (bears and other mammals) or changing the color of their coats (weasels, arctic foxes, and hares). For plants, photoperiodism causes leaves to change color in the fall. Across the Midwest, millions of acres of corn and soybeans grow, and the annual fall harvest is possible only when the plants senesce—triggered by changing day length—and mature their golden harvest.

The relationship between the gingko tree and leaf and the soybean photoperiod that I observed is purely a human-induced physiological phenomenon. While flying above a county road that was adjacent to a soybean field in late fall, I photographed an interesting pattern along the edge of a soybean field. Laid out in precise fashion was a series of green patches of nonsenesced soybeans, almost exactly mimicking the shapes of gingko leaves! My initial response was curiosity, so I enticed the pilot to fly closer, revealing the causative factor. A row of very bright streetlamps bordered the road, and their illumination each night was bright enough to alter the photoperiodic response of the soybeans. The nearby soybeans had not experienced night. The gingko shape was achieved by the power of the streetlight diminishing in an arc as it projected onto the plants. The split that created the two lobes was caused by the shadow of the pole. The effect, though, was quite striking—a perfect gingko-leaf pattern visible only from the air!

GREEN FEATHERS

MICHAEL JEFFORDS

The day turned out to be distinctive. The early-spring greenery of the cypress at Heron Pond (Cache River State Natural Area in southern Illinois) was alive with spring migrants—blue-gray gnatcatchers, prothonotary warblers, and the occasional northern parula warbler. These visitors certainly made the day distinctive, but not unique. That honor was reserved for feathers of a different color—chartreuse, celadon, and lime—colors associated with no southern Illinois bird familiar to me. These feathers were made not of keratin but of cellulose and chlorophyll and had migrated to the quiet waters of the swamp in unbelievable numbers. The American featherfoil (*Hottonia inflata*), a plant that we considered rare and had actively sought for many years, had arrived. Featherfoil plants existed by the thousands, even tens of thousands, on the sunny margins of the swamp. The round, symmetrical, green inverted candelabras intertwined in an ephemeral yet impenetrable lattice on the tea-colored water. We couldn't imagine where they had come from. The mere fact that they existed—bright-green asterisks running across the pages of the swamp—was reward enough for us.

Before our hike to Heron Pond on this sunny, warm day, I had debated the merits of lugging my usual assortment of camera gear. Rationalizing that I would likely see nothing new, I carried only a camera body with a wide-angle lens. Upon seeing all the featherfoil, I was torn between hiking the two-mile round-trip to retrieve additional gear or just using what I had. Laziness won out, but by getting low and allowing my feet and knees to be immersed in the swamp (ample punishment for a laissez-faire attitude), I was able to capture this most unusual event.

What circumstances had brought featherfoil to Heron Pond? At work on the following Monday, I consulted with a few botanical colleagues to find an answer. American featherfoil is a denizen of quiet swamps and a winter annual (a plant that germinates in the fall and winter and grows actively in spring). Interestingly, the seeds can remain dormant for many years, waiting for the right conditions to occur. Most swamps in the eastern United States dry to mud during part of the year, allowing the seeds of American featherfoil to germinate. Over many years, various beaver dams had kept the water at Heron Pond at an artificially high level, thus inhibiting the featherfoil's yearly cycle. However, during the previous year, southern Illinois experienced an extreme drought, and the swamp had dried out. Beavers were also removed and their dams destroyed. The opportunistic featherfoil took full advantage of the conditions and now covered the swamp in a festival of spring green. While American featherfoil still makes an appearance now and again, we have not seen such a display since, as the beavers are back and the drought cycle has dissipated.

American featherfoil (*Hottonia inflata*) appears each spring, but in different locations across the Cache River State Natural Area, Illinois (*right*). The plant is a primrose and related to shooting stars (*inset*).

PAZ TALE

SUSAN POST

Refugio Paz de las Aves, Ecuador, 2006—5:30 a.m. "Hurry, hurry! We're late!" shouts our guide. In the dark, I scurry down a slick, muddy path through an orchard where unseen tree tomatoes whack me in the head. Two days earlier, our guide asked if we would be interested in viewing antpittas. We collectively replied, "Yes!" Antpittas are round-bodied, short-necked, and long-legged tropical birds attired in colors that blend into the forest floor. They bound or hop through the understory. *The Birds of Ecuador* states "Shy and elusive, members of this family are all too infrequently seen." Our guide had made a phone call about a visit, and here we are at a private reserve where Angel Paz introduces himself and explains what we will see, all in Spanish. My adventure with Angel Paz, the antpitta man, has begun.

A selection of the seldom-seen birds at Paz Refuge: cloud forest pygmy owl (*top left*), dark-backed wood quail (*top right*), ochre-breasted antpitta (*middle right*), and toucan barbet (*bottom left*), and black-chinned mountain tanager (*bottom right*). Sue is all smiles after a tour with the Paz brothers (*center*).

On the steep trail, he cups his hands over his mouth, producing a vibrating trill that sounds like the whirring of wings. He then calls, "Manuel, Manuel, Maria, Maria, Vincha, Vincha," followed by more trilling. I hear nothing but water dripping off leaves. He calls again, tossing sticks and pebbles into the woods. Much to my surprise, Manuel, a giant antpitta, appears—long legs with a chestnut breast. Manuel stuffs his mouth with the worms tossed to him and hops away to eat unobserved. Maria soon appears; she too is a giant antpitta and more brazen—she follows us long after the worms are gone. The calling ritual was repeated for yellow-breasted and mustached antpittas, but only the first one appeared.

Later, during tea and homemade empanadas, Angel tells how he came to his "talent." It was while clearing a trail to a cock-of-the-rock lek on his property that he noticed a large, plump bird in the shadows. He tossed the bird a worm, and it hopped forward and swallowed the treat. Eventually, Paz won the trust of this shy bird, along with several others. With the antpittas associating Paz with earthworm snacks, they began to come when called, and he had a newfound business.

We visited again in 2013. Angel Paz and his brother led us down a trail carrying a fruit-filled bag. As we headed to a feeding station, Angel motioned to a nearby tree, where a pair of cloud-forest pygmy owls sat, side by side. Suddenly, the male was on top of the female. Had we just witnessed a mating? Angel motioned us to sit, and he began to cut fruit at the feeding station. One minute there was nothing, and then colorful tanagers, toucanets, and barbets were on Angel's head, at the station, or in adjacent branches! Fruit gone, we left and followed Angel as he called and softly threw pebbles. Down the slope in the tangles appeared a mustached antpitta. Where was Angel's brother? That question was soon answered, as he came up the trail with a dark-backed wood quail and chick following him. Before we left to have tea and snacks, Angel called "Vincha, Vincha" and threw pebbles and worms. Soon, at eye level on a branch sat an ochre-breasted antpitta—an otherworldly adventure. Our experience in 2013 was different, yet the same.

We spent time with a man who cares about his land and the organisms in it and is proud to show off these feathered gems. It did not matter that we could communicate only through hand signals and gestures—our smiles said it all. "Muchas gracias, Angel Paz!"

OLYMPIC-SIZE PAN TRAP

MICHAEL JEFFORDS

For those of us who collect, or have collected, insects, a prize from the New World tropics is the peanut-headed bug (order Hemiptera, family Fulgoridae). This is a unique relative of the cicada. Remove the large, hollow "peanut" from the head and the bizarre wing coloration, and the resemblance to North American cicadas is evident. Peanut-headed bugs occur in tropical, damp lowland forests of Central and South America. On several trips to this region of the world, I had failed to find an example of this striking creature. During a visit to northwestern Ecuador, however, our group finally succeeded, but certainly not in the way I had pictured encountering this truly bizarre insect.

A peanut-headed bug (*top*) lies trapped in the film of a motel swimming pool. Eyespots are displayed (*bottom*) when the insect is disturbed.

We were visiting an area of northwestern Ecuador within cutover rain-forest habitat that had largely been converted to oil-palm plantations—an area not on the itinerary of most tour companies. We were staying at a small hotel on the edge of San Lorenzo. By North American standards, it would have received less than stellar ratings (maybe one or two stars). For three entomologists, however, the place was literally a paradise. The rooms enclosed a rather large swimming pool that was seldom used but was brightly lit at night. Even though this area was heavily agricultural, patches of lowland rain forest still existed in isolated spots, and these were the places we visited during our stay. After each long day in the hot, humid environs of the rain forest, we would spend the evening by the pool,

sipping chilled Fanta and diligently watching the quiet blue-green surface of the water. Why? Because it functioned as an Olympic-pool-size pan trap for insects! As part of our entomological research projects over the years, all of us had utilized shallow water-filled pans strategically placed in various agricultural fields to sample the insect fauna present. Insects, mostly small beetles and leafhoppers, would inadvertently land in the pan and become trapped by the surface film of the water. It was a quick and easy way to sample. We soon realized that the well-lit pool was functioning in the same way, but on a giant scale. As we were far from most tourist routes, the staff failed to diligently remove all the insects that ended up floating about, so we were in pan-trap paradise! All manner of insects, from giant grasshoppers to large moths and enormous beetles swam about, waiting to be rescued and photographed. The prize of the trip, though, was when Joe, our colleague on the trip, found a large peanut-headed bug swimming about early one morning. Here it was, the treasure of my youthful insect collection, alive and colored in all its entomological splendor. The bug was easily retrieved with the seldom-used pool skimmer and proved to be surprisingly docile when handled. These insects are plant feeders, sucking sap from trees and shrubs, so they have no potential to cause bodily harm. Their only defense is the formidable-appearing head and large eyespots on the hind wings that are flashed when the insect is disturbed. Transport of our insect trophy into a nearby rain-forest habitat gave us endless opportunities to photograph our treasure in its primary habitat. Even though the climate was very hot and humid, we never did test the cool waters of the pool, as it had other uses for us.

MIXED MARRIAGE?

MICHAEL JEFFORDS

Look too closely at nature, and things can get very complicated, very quickly. Not long ago, while working on a field guide to Illinois butterflies, I was photographing a male viceroy butterfly in Bonnie's Prairie, an Illinois nature preserve in the sand region of Illinois, when along came a female red-spotted purple. Both are commonly encountered butterflies in the Midwest. Several things are of interest with these two species: First, they are closely related (in the same genus, *Limenitis*), even though they have wildly different color patterns. Second, their color patterns have evolved through a process called mimicry. The "ancestral" color pattern for *Limenitis* is dark, with large white bands across the wings. The orange viceroy is a Müllerian mimic of the monarch butterfly—both species are unpalatable to their predators and possess a similar color pattern. This allows predators a simpler mind-set (fewer patterns to recognize) when learning what not to eat! Monarchs feed on milkweed and sequester heart poisons, and viceroys feed on willow and have bitter aspirin-like compounds in their bodies. The red-spotted purple, however, is a Batesian mimic of the toxic pipevine swallowtail—here a palatable species mimics the color pattern of its toxic "model" to avoid predators who generalize color patterns when deciding which prey are good to eat. The caterpillars

The ancestral color pattern of *Limenitis* (*above*). Pictured on the opposite page is a female introgressive hybrid red-spotted purple being courted by a normal male viceroy (*top*), a viceroy (*bottom left*), a typical hybrid between a viceroy and a red-spotted purple (*bottom center*), and a red-spotted purple (*bottom right*).

of the red-spotted purple feed on a variety of trees and shrubs and are perfectly edible as adults. Consequently, both the viceroy and the red-spotted purple have diverged from the ancestral color pattern, and both are mimics of other species.

If this were not complicated enough, rare hybrids are known to occur between the viceroy and the red-spotted purple. While this may not seem that unusual, as they are closely related, when looking at their very different color patterns, it appears odd they would still recognize each other as potential mates. At Bonnie's Prairie, I witnessed a male viceroy actively courting a female red-spotted purple, and she was responding to his attention! It was quite unusual to see this interaction happening in the field. Butterfly hybrids are usually noticed only in large collections of insects. I was excited to capture an image of the activity, and when I returned to the office and processed the digital file, I noticed another level of complexity. It seems that it wasn't just two different species about to get together when both butterflies landed in a nearby tree with obvious amorous intent. Additional genetic shenanigans by the female's parents—introgressive hybridization or introgression—was involved. The pattern of the red-spotted purple female was not the typical design, but represented the pattern of the offspring of a red-spotted purple/viceroy hybrid and a normal red-spotted purple. When two different species mate (interspecific hybridization), we call their offspring hybrids (technically F1 hybrids). When the F1 hybrid mates with a butterfly with the normal color pattern, we call this introgression, as the offspring are not strict hybrids, but a genetic blend. In short, I witnessed a very odd red-spotted purple (an introgressive hybrid) about to couple with a normal viceroy. Ultimately, the photo took much less time to take than did the explanation of this unique encounter.

MÉNAGE À TROIS

MICHAEL JEFFORDS

When we hear certain terms, such as ménage à trois, definite images come to mind. Not surprisingly, the natural world sports a rich variety of sexual novelty. In the insect world, such antics have a specific purpose. Among the beetles, the most species-rich group of animals on Earth, one can find a great diversity of sexual behaviors. Interestingly, a driving force behind many sexual behaviors is *sperm precedence*, that is, the last sperm deposited is used for fertilization of eggs by females. To a close observer, beetle threesomes are not that uncommon. To understand what is happening, we must further explore sperm precedence. But first, a little background on our participants.

A female soldier beetle has two males attempting to mate with her (*top*). In the sequence above, two males attempt to mount and mate with a female green tiger beetle (*bottom*). The event shown here lasted less than five seconds and happened so fast the flash did not have time to recycle for the last photo!

Soldier beetles in the insect family Cantharidae are commonly seen in home gardens and natural grasslands on flowers, where they are important pollinators. They greedily feed on pollen and move it around accidentally to affect pollination. We call this phenomenon "mess and soil" pollination, likely the first type of pollination to evolve when plants first developed flowers in the middle Cretaceous period. While feeding, a female soldier beetle is often approached and mounted by a mate-seeking male. If he successfully copulates, his sperm will fertilize the female's eggs. Many times, however, a second male will also mount the pair and attempt to dislodge the first male and mate with the female. This battle of genital wits often leads to a mental image of ménage à trois, but it is really a battle for genetic supremacy.

Another group, tiger beetles in the family Carabidae, are voracious predators, inhabiting a variety of habitats. One of the most readily observed is the common green tiger beetle of the eastern woodlands. Forest walks are often accompanied by these bright-green denizens flitting just ahead on the trail. Green tiger beetles are highly active, mobile predators, and a male approaching and attempting to mount a female tiger beetle can be a dicey affair. If done improperly, the female may treat the amorous male as nothing more than a snack. When successful, the first male to mount the female is sometimes challenged by a second male, attempting to dislodge him from the female. The result is a momentary glimpse of a bizarre "piggyback train" of tiny tiger beetles. This mating behavior is seldom witnessed, but also driven by sperm precedence. I have observed as many as four males attempt this on a single female! Even though the female may have mated multiple times, ultimately, it's the sperm from the last male that will fertilize her eggs. Thus, it is not a worthless endeavor to try to dislodge a couple in flagrante delicto, for there is much at stake—the incorporation or absence of an individual male's genes in the next generation of beetles.

LOST IVORY

MICHAEL JEFFORDS

In the introduction to this book, I mentioned that as a child I created my own small museum. Who knew that nearly fifty years later, I would be in the dubious position of helping dismantle a very charming yet aging natural history museum at the University of Illinois? The official term was that we were *deaccessioning* the aged collection, transferring most of it to the Illinois State Museum. A small portion was to go to the collections of the Illinois Natural History Survey, where I am employed.

A pair of the presumed extinct ivory-billed woodpeckers, collected in Sunshine County, Mississippi, in 1892, were rescued from obscurity. Ironically, this area is just across the river from the Cache River in Arkansas, the site where living ivorybills may have been sighted in 2004.

The week before the enterprise began, Sue and I were walking through the dark, dusty halls, trying to assess what would be useful to the survey and what we could use for displays in our new building. The arsenic-laced lioness did not seem appropriate, nor did the skeleton of a baby great blue whale. Too big! I looked lustfully at the juvenile American bison, residing in its own glass case... but that is another story. The intricate and beautiful wax sculptures of mushrooms seemed appropriate, as did the plaster models of various midwestern reptiles and amphibians, so these were tagged. We also earmarked several large taxidermy specimens of Illinois herons and egrets, an owl or two, and even the delicately intricate models of Illinois forest wildflowers. All of these treasures were created by craftspeople in the 1930s and were priceless works of art, at least to a pair of biologists. As we were leaving, we happened upon a small workroom in total disarray; it was a staging area for plaster repair and used for paint storage. Thus, the brittle paint-spattered drop cloth in the corner was no surprise. On a whim, Sue picked up the drop cloth, just to see what might be hidden there. We were both stunned speechless when, from under the cloth, feathers ruffled from their less than "archival" storage, we saw the pair of ivory-billed woodpeckers pictured at right! What a treasure. Upon closer inspection, they were in surprisingly good shape, and we carefully placed the pair in a box for transport to our laboratory. The bills were no longer ivory (they turn a dull orange after death), but otherwise the colors were still vibrant.

Later, we discovered an accession number (written in pencil) on the bottom of the mount. Could this be the key to finding out more about these unique specimens? Paradoxically, the only part of the museum the university insisted on keeping was the large leather-bound catalog of all the museum's forgotten artifacts. It was placed in the University Archives. Armed with the accession number, one day I arranged to see this tome, which could be viewed only on-site. With a camera and a notepad, I went in search of the provenance of our unlikely pair. Thank goodness for the systematic nature of museum folks, for in the dusty yellowed pages I found out that these two extinct creatures had been collected at Long Lake, Sunshine County, Mississippi, in 1892 by a W. A. Stewart. What happened to the ivory bills? They are now on loan to the Barkhausen Wetlands Center, in the Cache River State Natural Area, the last place where this bird was recorded from Illinois. At long last, the pair has found a fitting home.

OTTIE AND BOKKIE

SUSAN POST

We are up early at the overlook at Shamvura Camp in northeastern Namibia, across the border from Angola. The camp lies on an ancient sand dune, one of the few elevated areas along the Okavango River—a sublime papyrus-lined stream. An African fish eagle lands in a nearby tree. In the distance, we hear the hollow call of a coppery-tailed coucal, when Michael suddenly calls out to our traveling companion, "Hey, Luann, why don't you have a rollick with Bokkie!" As for myself, I try not to look Bokkie—a large goat—in the eye; otherwise, I'll also be in for a rollick.

Michael is caught in "midrollick" with Bokkie, the goat (top photo by Robert Wiedenmann). Ottie enjoys sunning on the lawn after a day spent "in the field," defending his territory along the Okavango River.

Owners Mark and his wife, Charlie, share their lodge not only with guests, but also with a Cape clawless otter—Ottie—and Bokkie. Bokkie will sit patiently in a chair in the lodge's living room, but prefers a good battle. A "rollick" is wrestling, taking the goat by the horns, and shoving, pushing, pulling, and turning, often in pure self-defense, as the horns are on the same level as Michael's groin. Ottie the otter goes down to the river daily to mingle with his own kind, but reappears each morning to be fed a bottle. The third member of the household is Thug, a Jack Russell terrier, who likes to frolic with both the goat and the otter. Mark, a retired game ranger with many years of conservation experience, and Charlie, a midwife, are well known, respected in the community, and eager to share their love of the area. A community member brought Ottie to Mark as an abandoned pup. To have the opportunity to examine and interact with a Cape clawless otter was incredible. He liked to be petted and held, and I discovered his fur to be soft and silky. Instead of paws, he has strong nailless fingers and toes on his partially webbed feet. When we departed, Ottie came to our cabin to see us off.

In addition to Ottie, locals are constantly bringing new finds to Mark. The lodge's living room has evolved into a mini museum, with treasures that include drawers of local insects, mussels, and bird skins. One morning during our stay, I noticed the swimming pool had a surface film of feathers; what had happened overnight? Mark explained that a local tribesman found a dead spotted eagle owl and had brought it to him. Mark was holding the owl out to examine it when Ottie rushed him, grabbed the owl, and headed to the pool. With feathers flying, Ottie devoured the owl in the pool. The only evidence left were the talons at the pool's bottom and the floating feathers. Ottie was in hiding and would not even come for his bottle!

At the commencement of dinner, Bokkie was banished outside. Yet during meals, he would creep into the house and start edging ever closer. He would soon appear with head on table and nose almost in my plate!

As part of our stay, we are treated to one of Mark's signature bird outings, where he is literally on point. He stops and points his finger at a tree full of starlings that clothe it like leaves and then abruptly fly to the ground. Nearby, reed beds are laced with balancing little bee-eaters. More than once, Luann and I are admonished to stop talking and pay attention. As the sun sinks below the horizon, it is time to head to the living room, recap the day, and enjoy a sundowner—and, more likely than not, someone must have a good goat rollick!

TOY SALAMANDER

SUSAN POST

One of my favorite things to photograph is spring wildflowers, and while Illinois prairie groves can provide hours of entertainment, the wildflower assemblages of the Appalachians in Tennessee are sublime. Wave after wave of trilliums, violets, and orchids test plant-identification skills and tax my photo imagination. While most people think of Great Smoky Mountain National Park as the destination, I prefer the more intimate setting of Frozen Head State Park, near Wartburg, Tennessee. Frozen Head is located in the Cumberland Mountain range, and while the trails are rocky and steep, they provide one of the country's best spring wildflower shows. Wildflowers, however, are just one of the many surprises the park has to offer—the calls of black-throated green warblers are a constant along trails. On rainy nights, the pavement may be covered with hundreds of frogs on their way to ephemeral pools. Often our visit is timed to the emergence of tiger-swallowtail butterflies, flitting across our path as if in a Disney movie. Spring is wet, and most hikes require a rain suit for comfort. Yet it is this moisture that provides the park's magic.

On the Panther Branch Trail—no panthers yet sighted—plenty of wildflower and insects are present. My first experience on the trail has kept me coming

A red eft leaped from a nearby stump and landed on the trillium. They are very active and can cover quite some distance on rainy days.

back. It had just stopped raining—wildflowers look their best in soft gray light—and I was headed up the trail. Michael and I were busy with dwarf-crested iris and the various trillium species, and we soon separated, each seeking our own photographic zen. I looked on the ground and spotted a bright-orange salamander. I called to Michael, "Come here and see this. Some kid has dropped his toy." When Michael came, he just laughed and said, "That's a red eft, and it's no toy!" Really! A salamander exists with that color pattern? Apparently so, as red efts are one stage in the complex life cycle of the eastern red-spotted newt. The newt lays its eggs on plant stems in water, and when the eggs hatch the tadpole-like creatures—with legs—live in the water for two to four months. They develop lungs, move onto land, and transform into the orange-red creature I encountered. For two to five years, the eft leads a terrestrial life, wandering about the forest floor. Large numbers may be found after heavy rains, and their conspicuous bright colors warn predators that they are toxic. Their skin contains a secretion—tetrodotoxin—that protects them from most predators. The wandering red-eft stage helps the species colonize new aquatic habitats. The mature eft eventually returns to the water and transforms into an adult, and its bright colors disappear. The adult newt is dull brown, with a yellow underside.

After that first sighting of the bright-orange eft, sequestered among the ephemeral spring wildflowers, eft seeking has become a spring ritual. After tax time, you will often find us—rain gear donned—seeking our annual "fix" of toy salamanders.

ENDEMIC FAMILY

MICHAEL JEFFORDS

Groucho Marx's signature line from his 1950s TV show was "Say the secret word, and win a hundred dollars," and to a birder that secret word is *endemic.* A simple definition for endemic is "characteristic of or prevalent in a particular field, area, or environment." Advertisements for birding trips frequently emphasize this as a primary selling point—clients will "see *X* number of endemics!"

Male and female Yucatán jays (*Cyanocorax yucatanicus*) (*top*) land on a dead tree, followed by their immediate family (*bottom*). Note the yellow beaks of the juveniles.

Our quest for birds in the Yucatán Peninsula of Mexico included a long list of endemic "wanna sees," and the Yucatán jay was on the top of my list. Related to the familiar blue jay, this black and blue beauty is found only in the Yucatán and adjacent areas of Belize and Guatemala. We had traveled widely throughout the peninsula, visiting moist forests in the South, coastal marshes in the North, the vast Mayan ruins of Calakmul, and even the wetlands around Campeche, without seeing this prize.

When we finally encountered a family group, it was in the northern thorn forests south of Río Lagartos. It was early morning with a gray sky, and as we were nearing the end of our walk, a single Yucatán jay popped into view high in a dead tree. Our guide, Ishmael, immediately identified it as an adult, based on its plumage and black beak. He held up a finger and motioned us to silence, as he knew more was about to happen. Soon a second adult appeared, followed by three juveniles with bright-yellow beaks. They posed above me like stuffed birds in a museum display. In fact, several of my colleagues commented that the photos looked like I had taken them in the local natural history museum!

The Yucatán jay is a cooperative breeder. The young remain for a season and help the adults rear the next generation of chicks. The birds lingered above us for quite some time, with the juveniles looking longingly at the adults for a handout. Perhaps they were not yet old enough to help with this year's brood, or maybe these were newly fledged and were not quite indoctrinated into the ways of their world. These birds, like all jays, are omnivores and will feed on anything from fruits and nuts to lizards and small snakes.

For an endemic species (the term is often also associated with rarity), the Yucatán jay population appears to be increasing. A search of the literature reveals that in addition to the thorn forest, the species thrives in heavily degraded forests and even in hillside plantations. The morning's hike was very successful, as we encountered several colonies of army ants crossing the trail and experienced the steely glares of a group of ring-ready fighting bulls in a nearby pasture, but by far the highlight of the morning was our brief but fruitful encounter with ornithological endemism.

A CHANGE OF HEART

MICHAEL JEFFORDS

I t's not often that a place can so stun me that suddenly pictures are just not enough. My finger froze on the shutter and instead craved the feel of a pencil, scribbling away in a notebook. It was Colditz Cove in the Cumberland Mountains of Tennessee that inspired my initial venture into journal writing.

The trail, an old road through a pine plantation, held little promise. We had been on a hundred like it before. Soon the road narrowed, dipped, and became a soft, carpeted path leading through small to medium-size native hemlocks. Not far ahead, abruptly, we came upon a moss-carpeted rock ledge, and then, nothing. Nothing, that is, if you call massive-boled, arrow-straight hemlocks, three feet in diameter sixty feet up their trunks nothing! A huge circular bowl, geologists would call it a cove or cirque, spread out below. The floor was an uneven thicket of rhododendrons, pierced only by the huge hemlocks. To the right, as if the hemlocks were not enough, a wide, quiet stream also reached the abrupt edge of the sandstone precipice, and gently plunged, if water can be said to gently plunge, onto a soft, corrugated,

Northrup Falls drops sixty feet into Big Branch Creek in a stunningly beautiful cascade (*right*). A visit one December showed little change (hemlocks are evergreens), except the mist from the water had encased the foliage of the valley floor in an icy sheath (*inset*).

stair-step of Pennsylvanian-age sandstone. The falls were wide and thin, almost like a lace curtain. The trail led left and wound down into the green-clad valley. My thoughts drifted back in time to the obvious turbulent, rushing waters that must have carved this remarkable cirque. But from where did such a torrent originate? No glacial meltwaters reached here. Within the cove resided a series of huge shelter caves, undercutting the overhanging ledge. Giant chunks of sandstone, each toppled from the many-layered roof, lay in silent turmoil on the bright white eroded sand. Some appeared to have fallen recently enough that we dared not tarry long underneath. Immense as they were from above, the hemlocks defied description from ground level. Trunks were sunk deeply into clusters of boulders; some tilted at crazy angles. Around each ancient tree the moldy, black, wet humus that passed for soil gave all the appearance of fecund solidity. Here was a place, cliché or not, that time had literally forgotten. I felt that we might have been the first humans to see this place—a deep, cool window into the primitive past.

We tarried here for an afternoon, until an approaching thunderstorm threatened to reprise the cove's genesis. Back in the car, with photographs "in the can," I felt there was something left to be said. Not a feeling I was used to. When I asked Sue for a pen and paper, she looked at me quizzically. Journal writing was her thing, not mine. After a couple of hurriedly scribbled pages, the Muses moved on, and so did we.

TRAVEL HAZARDS

MICHAEL JEFFORDS

Before each trip abroad, Sue and I check in with our local Travel Clinic. Here we receive information on various medical "travel hazards" associated with the landscapes for our next adventure. Most requirements are fairly routine—tetanus, measles, and yellow-fever vaccinations. Other drugs are more curative in nature, should we inadvertently "drink the water." A final preventative usually involves antimalarial drugs. When preparing for a lengthy trip to the Peruvian Amazon, our travel company sent along a brochure about preparedness for this hot, humid area, including what to pack—appropriate clothes and photo equipment. A short sentence at the end of the brochure mentioned that this was an endemic zone for the "dreaded" tropical disease (the brochure left out "dreaded") leishmaniasis, transmitted by sand flies. They encouraged that we always have, and use, insect repellent that contained DEET. By using this repellent, our chances of contracting this disease would be minuscule. I intensely dislike being coated with chemicals in hot, humid weather, so I chose to ignore the insect repellent and instead covered myself with lightweight clothing and wore an insect-repellent hat. In addition, I thought that just walking in the wake of my DEET-drenched colleagues would protect me. Coincidentally, I was not bothered by mosquitoes or other biting insects during the trip.

I returned home happy and assumed that all was

Michael is clad in an insect-repellent jungle hat in the Peruvian rain forest. Given the outcome, perhaps a thin coating of DEET should have complemented the outfit!

well. Two months later, I developed a lesion on my left knee that simply refused to heal, and I had lumpy lymph nodes in my groin. A trip to my family doctor yielded a puzzled look from him, and I subsequently was sent to the local cancer clinic for screening. Oh, joy! After a nervous two weeks, the diagnosis was no, it's not cancer, but the oncologist was mystified about the lesion. By this time, my family doctor had retired, and I had a new physician. During my first "meet-and-greet" session, I pointed out the lesion on my knee. Before the visit, I had conducted my own research and concluded that I had contracted cutaneous leishmaniasis, caused by a protozoan parasite called a trypanosome. I asked my new doctor to order a test for this, but he was extremely skeptical and asked if I was just "picking at the wound." My response perhaps disconcerted him. I inquired as to how many parasitology classes he had taken during medical school (none), so I chimed in that I had taken four college classes, including protozoology. Hmmm.

To his credit, he did order a skin biopsy, but only reluctantly. After all, what were the chances? Three days after the test, I was informed that I did, indeed, have leishmaniasis. Back I went to the Travel Clinic, this time as a full-fledged patient in need of treatment. Given that the disease is so rare in the United States—I was one of only eighty-eight cases in the United States that year—the Centers for Disease Control in Atlanta had to be contacted because my local hospital did not have the necessary drugs to treat me and they were difficult to acquire. One month later, after nine all-day treatments with a potent antibiotic, the lesion healed, leaving a lovely scar—a quiet reminder to always read and abide by those short admonitions in your travel brochure, or suffer the consequences.

LEARNING FROM LEOPOLD

SUSAN POST

I first learned of Aldo Leopold during my sophomore year in college. *A Sand County Almanac* was assigned reading for my biology course. I read the book in the hospital while recovering from an emergency appendectomy. I thought, "How could something written in the mid-1930s and '40s have any relevance today?" It was just words about one man's observations on geese, chickadees, marshes, and old oaks. I read it only to answer questions that might appear on the final exam.

The famous "Shack" of *A Sand County Almanac* nestles in the restored landscape of an old sand farm (*top*). Sandhill cranes fly to the river each evening in late fall (*bottom*).

Once I completed biology class, Leopold and his book slipped from my consciousness, until I started writing for the *Illinois Steward* magazine. *A Sand County Almanac* was almost a required textbook for the magazine staff. The managing editor strongly believed in Leopold's land ethic, and those stewardship values helped shape the magazine. During my second or third reading of the book, its content took hold. I discovered not only Leopold's timeless conservation and land ethic, but also his elegant use of language. These were not just words, but illustrated classic literary devices—metaphor, simile, alliteration, assonance, onomatopoeia, and hyperbole. Leopold used all these to skillfully add "music" to his writing. With this newfound enthusiasm, it became required reading in the communication classes I taught for the public. I emphasized that one of the best ways to learn to write was to read the writings of others. Here was a concrete example of how to use the literary devices largely glossed over in rhetoric classes.

My discovery of Leopold the conservationist was not confined to multiple readings of *A Sand County Almanac*. Our magazine editor's obsession with all things Leopold led to an invitation for the *Steward* staff to spend the night in Leopold's famed Wisconsin Shack (a chicken coop that he refurbished for his family to stay in while restoring the land), walk the grounds where most of the observations for *A Sand County Almanac* occurred, and meet Nina, his daughter. What was staying at the Shack like? I saw the monument to the "old oak," watched geese fly into the marsh, and observed chickadees bouncing among the pines. The one thing Aldo Leopold was not proficient at, however, was building a fireplace! It was not vented properly, and the smoke-filled room did not lend itself to a restful night. In my journal, I created a simile for my stay: "I awoke smelling like a smoked Smithfield ham!"

One of the questions posed to Nina during our visit was how her father wrote such elegant prose. She explained that he had a talent for it, and the words just flowed from his pencil. Upon returning to Champaign, I dug a little deeper and found original technical articles that were the basis for the later essays. Later, one of the many writing classes I taught was for the University of Wisconsin Arboretum. Here, I was introduced to the Leopold library, where I could see his original essays, typed and marked up. Apparently, he struggled with writing just like the rest of us.

Our visit to the Shack concluded around a campfire as we watched numerous sandhill cranes, like so many ancient pterodactyls, fall from a chilled, darkening fall sky. That vision I will not soon forget.

BE PREPARED

*Unfortunately there seems to be more
opportunity out there than ability. ...[W]e
should remember that good fortune often
happens when opportunity meets with
preparation.*

—Thomas A. Edison

For our honeymoon, we traveled to the northern Lower Peninsula of Michigan to find and photograph orchids. We camped and took only the bare minimum (a tent and sleeping bags). We scoffed at having a stove and planned to eat yogurt and cereal for breakfast and cold hot-dog sandwiches for other meals. Obviously, we were on a tight budget. Our only other experience "couple camping" had been in Florida, where a stove and hot food were not warranted. Mid-June in Michigan can be chilly, and a cold breakfast or an uncooked hot dog quickly proved to be not that appealing. Fortunately, we found plenty of orchids to photograph, but ended up spending most of our limited money on hot meals. Had we done any "homework," we might have prepared for Michigan's fickle early-summer weather and realized that camping in Michigan is not the same as Florida.

Both images—the pair of lilac-breasted rollers in South Africa (*previous page*) and the metalmark butterfly in Ecuador (*left*)—were created using the same 300mm telephoto lens. Sometimes simpler is better when it comes to being prepared with the right equipment!

We still camp, but now have thick thermal pads, a rain canopy that is the envy of most in the campgrounds (the result of too many missed meals due to rainstorms), lightweight cookware, and a reliable stove. The popular phrase *Be prepared* is now second nature to us, both as field biologists and as nature photographers. Lacking preparedness in those disciplines can lead to missing crucial data points for a research project or failing to photograph that "decisive moment." Nature is incredibly diverse and seldom predictable. As photographers, we know that anyone going to a specific site, such as Yellowstone, can come away with awesome images. However, how many were prepared to capture the moment when bison briefly materialized from thick fog only to disappear an instant later or capture a mother otter weaning her young by playing with a freshly caught trout as they hungrily watched?

In nature photography, when we go afield, seeking to document both the common and the bizarre, we make decisions beforehand that can determine our "preparedness"—for example, what equipment to lug along. Is this excursion for birds (long, heavy telephoto lenses), scenery (wide-angle lenses), or close-up work (macro lenses with accessory flash units)? Are there low-light conditions that will require a tripod? These decisions add up to preparedness. However, practicality must enter into the equation, as it is seldom feasible to drag along forty pounds of camera equipment to "do it all." Add hybrid trips to the mix that are not strictly for photography yet have the potential for unusual opportunities, and the whole preparedness process can become bewildering. Being prepared is not so easy, and missed opportunities weigh upon our collective psyches for years to come.

Another layer of preparation is to educate ourselves about any new area we are going to visit. If it's South Africa, an astonishing array of unfamiliar creatures will be popping in and out of view, and the question of "What is it?" repeated a hundred times a day weighs on even the most patient of guides! For our trips, we divide up the responsibilities by focusing on different aspects. Sue diligently studies the wildlife, even going so far as to create her own field guides. Michael familiarizes himself with the landscapes and is responsible for handling all the images created during each trip. Both of us carry different camera equipment that allows us greater flexibility—Sue's shorter telephoto lenses are suitable for everything from lilac-breasted rollers on the African savanna to tiny metalmark butterflies in the rain forests of Ecuador. Michael's longer telephoto lens can bring distant birds into sharp focus, while his macro lenses depict an unseen world.

In this chapter, we feature several essays that provided unique challenges, either because they were totally unpredictable, and even somewhat frightening (see "Beautiful Musk"), or just plain difficult due to the circumstances associated with the event (trying to photograph a cock-of-the-rock lek in near total darkness—see "Cock-of-the-Rock"). The good news is that we are always open to new challenges and try to take advantage of opportunities that are presented. It's a never-ending adventure that makes life worth living.

COCK-OF-THE-ROCK

SUSAN POST

Cock-of-the-Rock Lodge, August 16, 2010—Michael and I are up early, bustling about with headlamps—the lodge on the downslope of the Peruvian Andes where we are staying has no electricity, and it is hard to adjust to using only candles. We stumble to the small bus, heading for an early appointment with a cock-of-the-rock lek. The lek at our lodge was destroyed by a rock slide, so we must travel to another. From lodge to lek is just a few minutes trip. Yet in the dark, it seems like forever. By 5:20 a.m. we are at the lek, and all is dark and silent. Our bus driver leaves us (probably to catch up on sleep), and, after a short traverse down a steep, muddy trail, we wait. I can't even see to write, and all I hear is rushing water. Unseen birds announce their presence, but without our guide José's translation, all of the birdsong is a foreign language that I hear, but do not understand. At 5:35 a.m. I hear a loud "whoak," soon repeated by others. A cicada-like buzz accompanies the *whoaks*. It is still too dark, so nothing but noise arouses my senses. Finally, my eyes somewhat adjusted, I spy a large male bird bobbing and dipping, flaunting its silver wing patches. Other males soon join in. I notice a female, light brown with blue eyes—she forms the apex of a triangle while a line of three males vies for her attention. At 6:15 she leaves, but I'm not sure if the males are aware of this. By 6:25 a.m. there is nothing but silence, and we, too, depart.

One of the birds we sought during a Peru birding trip was the Andean cock-of-the-rock, the national bird of Peru. We didn't want to just see the bird; we hoped to experience its lekking behavior, called by some "one of the most amazing spectacles in the animal world." A lek forms when males gather to posture and perform competitive displays to entice mate-seeking females to approach. Our lodge had the largest known lek in the world, until heavy rains and landslides in 2010 destroyed it. These were the same storms that wreaked havoc with the train tracks leading to Machu Picchu.

Cock-of-the-rocks are forest birds that show a preference for steep-sided ravines with fast-flowing streams passing through them. While the females blend with their surroundings, the males are a flamboyant red-orange, the color of the setting sun during a hot summer evening. In the darkened forest, they try to grab the attention of females by flaunting their gaudy garb. The males perch on branches and vines six feet or higher above the ground; each male has his favorite branch on which to perform. And what a show it is! The leks are a confusion of noise (caterwauling comes to mind)—bowing, bobbing, jumping, flapping of wings, displaying the gray flight feathers, and clacking of beaks. Were any of these avian antics successful? During our fleeting visit, we saw no mating. In fact, the female stayed at the lek only briefly, seemingly unimpressed. We, however, headed back to the bus with goofy grins on our faces, spellbound by the spectacular performance!

A pair of male Andean cock-of-the-rocks pause in their antics to see if a nearby female is interested (*top*). The female briefly took note, but then quickly flew away, obviously unimpressed by the display (*bottom*).

VANESSA!

SUSAN POST

I am birding the maze of trails at Amazonia Lodge in Manu Biosphere Reserve, Peru. Our guide, José, pulls ornithological magic (with bionic eyes and ears) from the adjacent tangles and thickets. Yet even I can't keep focused on the birds flitting about when everywhere are butterflies. I pause now and then for a quick photo, but do not tarry; I need to keep up with the group. José points out tapir tracks in the mud as we linger to listen to a black-tailed trogon calling above the din of cicadas. I try to take a photo of mating translucent tan butterflies and step off the trail for a better angle. Suddenly, I hear a crashing, thrashing, thundering noise...

Vanessa, a wild tapir, returns periodically to the lodge where she was raised. Note the fleshy nose that is distinctive of tapirs.

I had spooked a resting tapir—our group's eyes were wide in response, and none of us was game to venture off the trail again. Needless to say, the butterflies were long gone. José assured us that at the next lodge, if we wanted to see tapirs, we could schedule a night hike to a lick where tapirs could usually be found.

On the eight-hour boat journey to our next destination—Manu Wildlife Lodge—I took time to contemplate tapirs by reading from my "trusty" homemade field guide. Tapirs are "pony-size," ungainly looking ungulates and the heaviest land mammals in South America. They are almost hairless, with the exception of a dense, short mane of bristly hair on their upper neck that aids them in making their way through the nearly impenetrable tropical undergrowth. Odd-toed, they have a prominent central digit bearing most of their weight. Toe splaying prevents them from sinking into soft, muddy ground. Perhaps their most unusual characteristic is their short elephantine snout, made entirely of soft tissue. This ever-twitching appendage not only aids in sniffing out food, but can also be used as a snorkel. If a jaguar attacks, a tapir can jump into the nearest water and swim out of danger with only the tip of its nose visible. Tapirs rarely stray far from woodlands and are most frequently found close to water. The final entry in my booklet stated, "They are shy, silent, and rarely seen." Wow—to see and photograph a tapir after my recent experience would be something special.

We finally arrived at the lodge, hot, sweaty, and tired, just in time for dinner. I was ready to sign up for a three-hour tapir-viewing trek through the rain forest (at night) when suddenly the kitchen staff yelled, "Vanessa, Vanessa!" Is Vanessa a jungle superstar and this evening's entertainment? All the guests ran for the door, after first emptying the fruit bowl. Something was up that we were not privy to. Michael and I followed, wondering, "What is going on?" Vanessa turned out to be an adult tapir that had been raised by local people after losing her mother. After being released into the wild, she periodically returned to the lodge for fruity treats. She was tame enough that we could stroke her and take photos. She smelled earthy and felt muscular—like petting a hairy slab of granite. She appeared only that first night. We passed on going to the lick, as the quarry had come to us. What a treat!

BEAUTIFUL MUSK

MICHAEL JEFFORDS

While I was in graduate school, my professors worked diligently to prepare me for the life of a scientist, and they did a good job. What they failed in, however, was mentioning that the world we live in can be very strange and close encounters of the "bizarre kind" are not outside the realm of possibility. Such an encounter happened one summer day, just outside East St. Louis, on the flat plains of the American Bottom. I had spent all day sampling soybean fields for bean-leaf beetles and was headed home when I drove by a wheat field ready for harvest. The low afternoon light cast a beautiful glow, and I was struck by a lone thistle growing amid the wheat. The scene was quite striking, and I stopped my university vehicle—with the official state seal on the side—set up my tripod, and was busy photographing. I stopped only when I heard an ominous double click to my right. I am not a hunter, but I knew the sound of the hammers being drawn back on a double-barreled shotgun! A quick glance revealed a red-faced, overalled, and pissed-off farmer. While the gun was not exactly leveled at me, it was still intimidating. We locked eyes, and he stated, "What you doing here, boy?" I took a few moments to collect myself, before answering, "Just photographing this wheat field; it's really quite pretty." Those lame words had little effect on him, as he continued with, "You better put away

A lone musk thistle decorates a wheat field ready for harvest (*left*). Musk thistles demonstrate why they are an exotic invasive species by dominating an old field (*right*).

that camera and git." By then I had recovered a little from my shock and tried to explain what I was doing in the area. To his credit, he listened, but I'm not sure he really believed me.

I soon found out what the problem was, at least from his perspective. This was his wheat field, and even though I was not trespassing, he saw that I was focused on the thistle—which was the exotic, invasive musk thistle. The species had just been declared a noxious weed by the state of Illinois. All noxious weeds, by law, are to be controlled when possible. His perception was that I was in a state vehicle, and I was documenting the "fact" that he had failed to remove this noxious new weed. I assume he saw a fine in his future, but was having none of it. The fact that I was totally unaware of the weed scenario at the time was irrelevant to him. His final statement to me was, "Get on out of here, and, by the way, why don't you go on up to the state park and look at their grounds? They got more of them damn weeds than you can believe. Tell them to kill them before they come looking in my fields." With that the encounter ended, and he rumbled off in his truck that I had failed to hear drive up.

My curiosity piqued, I drove the few miles north to Horseshoe Lake State Park, and, lo and behold, Mr. Farmer's statement had been right on the money. At least two acres were completely inundated with musk thistle, making an impenetrable thorny thicket that was likely the seed source for most of the surrounding agricultural land. All of a sudden, his perception of the previous scene came into sharp focus. Mine, however, failed to change, as both scenes remained remarkably attractive, and I enjoyed photographing them both.

BISON MIGRATION

MICHAEL JEFFORDS

Until recently, any student in the natural history sciences attending the University of Illinois spent time in the historic Natural History Building. Its exterior resembles a set for a Harry Potter movie, and its interior was a treasure trove of natural history artifacts. The creaking wood floors were always alive with students, all basically oblivious to the cases full of geological wonders, the mammoth plesiosaur skeleton embedded in the wall, and even the iconic bull bison that greeted all who entered the building. As a graduate student, I spent many hours there and enjoyed, even marveled, at the wonders contained within.

A bull bison—removed from the University of Illinois's Natural History Museum on skateboards—arrives at the Illinois Natural History Survey, where it resides today.

Many years later, when the aged building was to be renovated, I became involved in the deaccessioning of the natural history museum. This was both a sad occurrence—the university would no longer have a natural history museum—and a fortunate one, as the institution where I work (the Illinois Natural History Survey) and the Illinois State Museum were to remove all specimens that each institution wanted. All of this, however, had to be accomplished within a week's time, so anything that was desired—and could fit through the narrow doors or be disassembled—was removed. It was quite an experience (see "Lost Ivory")!

Several weeks later, the bison we had passed over was still on my mind. Would it be destroyed when renovation began? In its enormous glass case, obviously constructed in situ, it would not fit through any door in the building. What to do? After a short consultation with our shop crew—Mac and Larry—and with permission from the university, off we went to explore the possibilities of moving this historic bison. The massive glass and oak case was not meant to be moved, so we disassembled it, leaving only the dusty bison on an ancient sheet of plywood. With a wooden interior framework and stuffed with arsenic-laced sawdust, the bison was heavy and unwieldy. Fortunately, it just fitted through the outside door (with an inch to spare), but the doorway was at the bottom of a staircase. Our ingenious shop staff fastened four skateboards to the platform and built a ramp down the stairs for the bison to none too gently slide from its pedestal on its way to see the sun for the first time in nearly a hundred years! The bison was strapped to a waiting flatbed trailer. The next problem involved what direction to move this large fragile cargo. The truck and trailer were too large to turn around in the narrow museum drive, so the best course of action was to simply drive our beastly burden directly down the sidewalks of the university quadrangle, past the stately Georgian buildings, to our shop. I rode along on the trailer to stabilize the load, and believe it or not, not a single student of the hundreds we passed looked up at this unlikely caravan or acknowledged that a male bison was migrating across the quad. Remarkable! Safely at our shop, and after some much-needed pelt repairs to the creature (shot and stuffed in 1870), it now proudly resides in the foyer of the Forbes Building, where it has finally gained the notoriety and attention it so richly deserves.

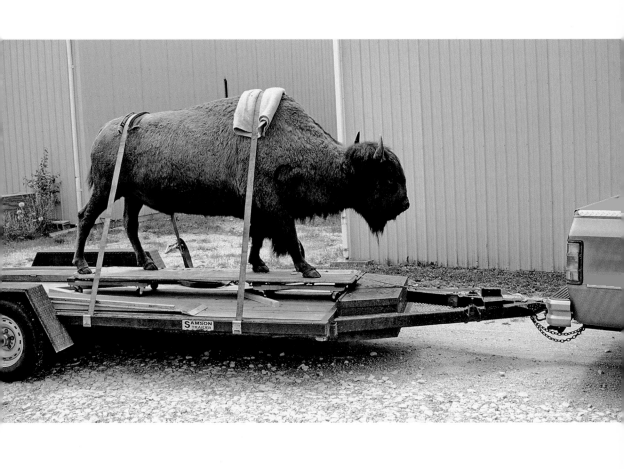

RAPACIOUS!

SUSAN POST

Our first trip to Yellowstone National Park was pretty much unscripted and turned out to be a bewildering and often frustrating experience. While we enjoyed ourselves, we were often wet, cold, and mostly unprepared for the wonders and rigors of the park. From that I learned to always prepare for a trip. If within the United States, I map out endless poten-

A red-billed oxpecker flies to a rhino (*top*). Oxpeckers explore Cape buffalo (*top right, middle*). Multiple oxpeckers gather on an aging giraffe that is rife with ticks (*bottom*).

tial trails and interesting sights as possibilities. Our time is precious. If it is an international trip, I study the organisms of the countries in question and construct my own special field guides. A second trip to South Africa was no different, and even though I had previously studied, it is Africa, after all, and I had time to learn only a small fraction about African wildlife.

As I read about spotted hyenas, the word *rapacious*—aggressively greedy or grasping—was used to describe their feeding. Hyenas are "rapacious eaters and will consume victims in their entirety." The word resurfaced in another field guide to describe the feeding of tawny eagles—"a rapacious predator." Obviously, *rapacious* was an important predator concept. I added it to my African vocabulary and eagerly anticipated using the word for a daily observation. A journal entry for Kruger National Park reads: "Today's word of the day—rapacious—applies to oxpeckers, both red- and yellow-billed. Over a dozen are clustered on a giraffe's neck, going about their business of grabbing and gobbling ticks—leading to another name for these

birds—tick bird. Up and down they venture, over neck, legs, and tick-covered hind end—a never-ending job!"

Oxpeckers epitomized my new word. They are considered to be parasites, are highly social, and are associated with animal gatherings, including Cape buffalo, rhinos, and giraffes. Oxpeckers travel with the herds—they feed, rest, preen, court, and even mate on the animals. The animals' fur and dung are used for nesting. The birds have curved needle-sharp claws and long graduated tails with stiff pointed feathers. The tail is used as a prop, much like a woodpecker's, while their bills rapidly nibble and pluck about in the hosting animal's fur. As they forage, oxpeckers cling closely to the fur, giving the illusion of legless birds. Oxpeckers clamber all over the head and body of their chosen animal with jerky woodpecker-like motions, searching for ectoparasites (ticks), wound tissue, and fluid secretions. Whether sedentary or slowly on the move, as long as the host is not running, each animal is given a thorough search. The birds even venture into nostrils and ears, inserting their head into the former and their whole body into the latter.

Our last morning on safari, we located a buffalo herd eating dry grass and moving slowly, like a great brown wave. A flock of oxpeckers flew in just as we ventured into the herd with our jeep. The birds, as if they had read the field guides, flew down from nearby trees and systematically explored each buffalo—ears, nostrils, and boss (area of fused horns)—over, under, around, and through. The hissing and churring of the birds, the swish of dry grass, and the grunts, snorts, and munching by the buffalo were just another African morning captured by two rapacious photographers. This time, we were prepared.

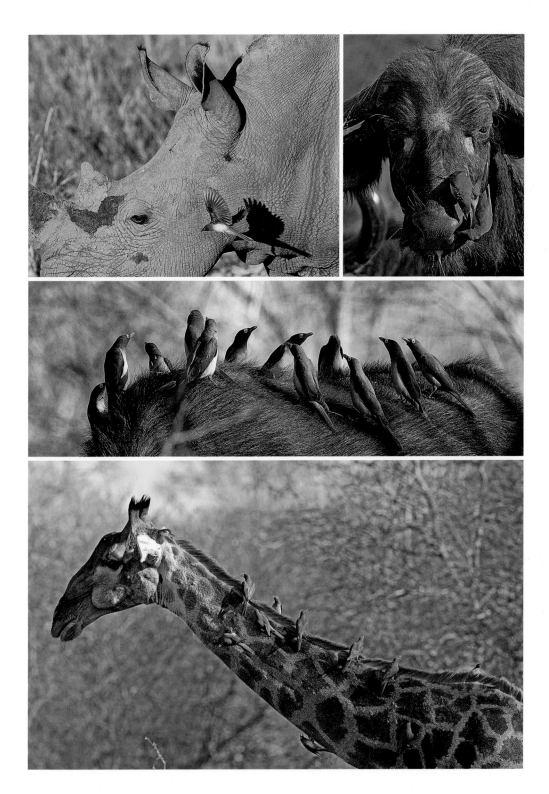

CONCLUSION

THE END OF THE LANE

SUSAN POST WITH MICHAEL JEFFORDS

It is eight o'clock on a sunny spring day, and I am in the Cache River wetlands once again. Today is Nature Fest, and as members of the planning committee and the Friends of the Cache River Watershed board of directors, Michael and I must be there early. Nature Fest is our annual spring event held at the Henry N. Barkhausen Wetlands Center. Its goal is to introduce citizens to the Cache. A variety of canoe trips, hikes, and exhibits are planned for the day, and each is to be staffed by a cadre of volunteers. My early-morning task is to make sure the volunteers park not in the Wetlands Center's parking lot—that is for attendees only—but in the grassy field just to the south. All volunteers must report in by nine to obtain their assignments, help with setup, and receive last-minute instructions. My assignment is to stand in the middle of the short access lane—the entrance to the Wetlands Center—dressed in my "official" lime-green Swamp Geek shirt. I am holding a giant arrow that emphatically points south to the volunteer parking lot. I wave my arrow about while I wait for cars to arrive, feeling a bit like one of those inflatable flying-tube guys usually flapping in the breeze in front of mattress sales and the like. Standing here also brings to mind another country lane where I caught a yellow school bus for twelve years. There was never time to ponder what was across the road or flying across the sky, as my siblings and I were usually making a mad dash to be at the end of the lane before the bus.

Today, I have plenty of time to take notice of my surroundings. To my right is an open-water cypress wetland hidden from view by a thicket of trees and shrubs; across the road is a grassy wetland, while on my left is a small pond, skirted by the newly mown area where volunteers are supposed to park. Michael is manning the volunteer entrance a quarter mile down the road, so periodically I wave. Blackflies buzz about

my head, and although I am alone there are plenty of distracting conversations around me. This is the height of avian migration and nesting in southern Illinois: common yellowthroat, prothonotary, and northern parula warblers; red-winged blackbirds; blue-gray gnatcatchers; mourning doves; and a lone Bell's vireo are calling and flitting back and forth. Hidden cricket frogs add their "clacking" to the dialogue, and silent tree swallows sally across the sky, hopefully eating these pesky flies. An adult bald eagle sweeps across the sky, its white head highlighted against the day's cloudless blue. A trio of little green herons squawk and fly across the road. Lately, I have had a hard time finding any of these cryptically colored birds, but today they are keeping me company with their comings and goings. In the distance, a large bird is approaching. Osprey! They are not nearly as common in the Cache as bald eagles. This osprey is intent on the small pond to my left; it circles and does a couple of fake dives before SPLASH! It exits the water with a fish grasped in its talons and flies off toward Michael, leaving nothing but ripples.

I continue to greet arriving volunteers, wave my sign, and provide directions. Several people will mention how they appreciated having someone stationed at the end of the lane to greet them. They just knew it was going to be a good day. In the tangles across the road, a hidden catbird periodically mews. As I try to locate the elusive gray wraith, a yellow-billed cuckoo pops into view. These birds are always secretive, but this one presents a perfect photo opportunity. Alas, I am holding a big goofy sign, not a camera. It is past nine, and as I look up and down the road one last time for final stragglers, a small flock of dark birds with white rumps melodically flies overhead—a favorite of mine, bobolinks!

Finally, it's time to saunter up the lane to the

Wetlands Center to begin my next assignment. Before going on, I see Michael approaching and excitedly ask him. "Did you see the eagle and the osprey? Did you hear all the birds?" Who knew standing at the end of the road could be so exciting? Yes, he had

seen the birds, and no he hadn't heard much, and he asked me why was my sign pointing up to the sky most of the time. The answer to that was easy: I was trying to write in my three-by-five-inch notebook. I didn't want to forget any of today's discoveries. Here was a place I had entered and exited many times, yet I had never taken advantage of its possibilities. My eyes were once again opened to the magic of discovery that can happen anywhere your curiosity takes you.

———

As Sue and I stood in our respective locales, flapping and gesticulating about to ensure that no one "violated" our self-imposed rules on parking, it occurred to me that this activity has mirrored much of our lives together. We have gone about our business, each in his or her own way, but have not been so focused on the task at hand that we abandoned our innate curiosity about what was going on around us. We are always open to "collecting" new observations. As Richard Conniff noted in his book *House of Lost Worlds*, "They wait for someone to whom a drawerful of specimens, silent for ages, may begin to resonate with lost voices and ancient dreams." So has our lifetime cabinet of observations accumulated and finally coalesced to lead us to produce this work. Each observation is subject to interpretation—ours alone where science has failed to articulate any explanation. I am struck by the fact that even if we are totally wrong about what we think is happening, it is less important than the observation itself—whether a photograph or just a notation in our notebooks. The observation is immutable, much like a colorful bird skin with dead cotton eyes lying in a museum cabinet or an ancient insect encased in a fragment of cloudy amber. Each awaits an inquiring mind to happen along and dispel the mysteries held within. This collection of observations and essays is quite literally our museum of the mind.

EPILOGUE

UNEXPECTED SURPRISES

SUSAN POST

When we travel, hike, explore, or even do field-work, we have goals and usually know what to expect. Yet our best experiences have been the unexpected—the surprises. Our intended target on a trip to the Pantanal in Brazil was to photograph hyacinth macaws and jaguars. We were not disappointed, but the highlight was spending time with a giant otter family—so close I could hear them crunch their fishy prey (see "To See a Giant Otter"). In the Falkland Islands, I knew we would see penguins, but had no idea I would encounter a "love-struck" teenage stalker chick (see "Stalker").

In between writing essays for this book and finishing last year's field notes, I have been studying for an upcoming trip to East Africa—a reward for both of us. I sit on the floor in my upstairs room, poring over the seven-volume set *Birds of Africa*. The books are oversized (twelve by nine and a half inches), quarto in library speak, and each one is more than five hundred pages. I am grateful that the University of Illinois library has the complete set and that I can borrow them at will. These are not books to lug along in a pack, as the set weighs upwards of fifty pounds! In scope, they cover the entire continent, and I peruse each tome, checking range maps (areas where the birds are) and comparing them with a current, but much smaller (and portable), *East Africa Field Guide*. This book will accompany me on the trip. While the information in my small field guide is adequate, the identification characters and general information sections in *Birds of Africa* are superb. Here, I often find quirky behaviors and foraging strategies; these tidbits are duly noted in the margins of my current field guide. They then become part of a mental wish list—I hope to see that!

This will be our third trip to Africa, and while I will see some birds that should be familiar by now, there are always new names and additional species to learn. Each time I read the Africa tomes, different birds seem to hold my attention. During my preparation for our second trip, I spent many hours reading about the myriad raptor species. One of these, the African harrier-hawk, or gymnogene, piqued my curiosity. Under the general-habits section of *Birds of Africa* were line drawings of the bird's unique feeding methods. Gymnogenes forage with a slow, systematic search; their long, slender legs can bend backward and sideways at the tarsal joint (they are double-jointed), permitting their insertion into awkward holes and cavities of trees and rocks. Its small, slim head can follow its claws into these narrow areas. These adaptations enable gymnogenes to pull out bats, lizards, and nesting birds. As I read this, I thought to myself, "Yeah, sure, I'll be likely to see that!" But my curiosity was roused, and that tidbit remained in the back of my mind. From my field notes on day two of our visit to Kruger National Park, I noted that our guide yelled, "Stop, stop! Juvenile gymnogene hunting on the right!" I looked up, and it was just like the line drawing in the book: the bird was probing a tree's cracks and crevices, twisting and turning, fluttering its wings. Suddenly, in its talons was a frog. Wow!

A few days ago, I was reading about bee-eaters, small colorful birds known for their aerial acrobatics during the pursuit of insects. I jotted down a few notes about the behavior of the white-fronted bee-eaters. These birds roost together in family groups, and many times several groups will perch together outside their roost holes. At dawn and dusk, all the birds gather, en masse, to perch, preen, and socialize for about an hour, before heading to their foraging grounds. There they separate into family units. This passage described a remarkable scene we had experienced during our first trip to Africa. Along the Okavango River, Michael captured photographs of behavior. We were actually there

to find Pel's fishing owl, a large cinnamon-colored bird the size of a cocker spaniel, that feeds on fish. While I was searching for the owl with my birding companions, Michael was near a riverbank, mesmerized by the white-fronted bee-eaters coming in to roost. His description of the experience is in the essay "Bee-eaters!" The next morning, up early to look for the owl again, our group spied hundreds of bee-eaters at the water's edge—basking, preening, and just hanging out. I hurried back for my camera, but they were all gone by the time I returned. Many years later, I now know what to expect and that I should've had my gear to begin with! That's a mistake I won't make again.

While I have my lists and homemade field guide of what we expect to see during this latest third trip, it is the surprises that I am most excited about. As Dylan Thomas wrote, "When I experience anything, I experience it as a thing and a word at the same time, both equally amazing." Our travels are like that. Prior to a journey, I research birds, animals, and the places—my imagination is filled with possibilities, names, and hopes. During the journey, with my homemade field guide as reference, I will fill numerous three-by-five-inch notebooks and photo flash cards with observations, attempting to record and remember everything. After the trip, it will be time to combine my notes with Michael's, as well as our thousands of photographs, to preserve the fleeting memories of a successful adventure. In short, I simply want to travel like Charles Darwin, "as a self-paced visitor with an inquiring mind."

A juvenile African harrier-hawk searches for prey (*above*). Sue prepares for an upcoming trip and photographs bison at Theodore Roosevelt National Park (*right*).

GLOSSARY

allomone—a chemical substance produced by an organism as a means of defense

anemophily—wind pollination; where pollen is transferred between plants by the wind

anoxic—the absence of oxygen

Anthropocene—the current geological age (recognized by some scientists), defined as the period during which human activity has been the dominant influence on Earth

apex predator—an animal at the top of the food chain on which no other creatures prey

aposematic coloration—conspicuous patterns or colors that advertise a warning to potential predators

Archaeopteryx—a group of dinosaurs that is transitional between feathered dinosaurs and modern birds

baleen—a filter feeding system, made of long bristles of keratin, found inside the mouths of many whales

Banksia—a genus containing around 170 species in the family Proteaceae, found mostly in Australia

barrens—a type of ecosystem with relatively infertile soil and few trees

basalt/basaltic column—a fine-grained, dark rock of volcanic origin that often has a columnar structure

Batesian mimicry—when an edible animal or plant is protected from predation by resembling an inedible organism

biogeography—the study of the natural distribution of organisms across Earth

biomass—organic matter that is derived from living or recently living organisms

black light—a lamp that emits mostly long-wave ultraviolet light

bog—a type of wetland that accumulates peat; dead, partially decayed plant material

brackish—slightly salty, as where freshwater marshes and wetlands meet the sea

broad spectrum—a pesticide that acts against a wide-range of insect pests, such as DDT

buttress—a type of support structure

cantharophily—pollination by beetles

carapace—the hard upper shell of a turtle

cardiac glycoside—a type of sugar that disrupts heart function by acting on the contractile force of heart muscles; a component of milkweeds

carnivorous plant—a plant that catches and digests insects to supplement its nutritional needs

chitin—a tough, fibrous polysaccharide that is the major component of the exoskeletons of Arthropods

chytridiomycosis—an infectious disease in amphibians, caused by the chytrid fungus *Batrachochytrium dendrobatidis*

ciliate—a group of one-celled animals (Protozoa) that have hairlike organelles called cilia

cirque—an amphitheater-like valley formed by water erosion, often glacial meltwaters

colic root—a white-flowered relative of lilies that prefers very sandy soil

conifer—a cone-bearing tree with needlelike or scaly leaves

continental shelf—an extension of a continent that is covered by the sea

coronal mass ejection—a huge burst of gas and magnetic field from the sun's corona that is released into the solar wind

cove—a recess or curved valley in the side of a mountain

DDT—the compound dichlorodiphenyltrichloroethane that is colorless, tasteless, nearly odorless, and known for its wide-ranging environmental properties; its use was banned in the United States in 1972

deaccession—to officially remove an object from a museum, library, or art gallery; it may then be sold or moved to another location

deciduous—a tree or shrub that sheds its leaves on an annual basis

decisive moment—a term applied by famed photographer Henri-Cartier Bresson to denote the climax or critical moment of an event that is captured by the camera

decomposer—any organism (bacteria, fungi, insects, and so on) that breaks down organic materials

DEET—a yellow oil that is the most common ingredient in most insect repellents

diapause—a period of suspended development in insects, usually in response to unfavorable environmental conditions

dieldrin—developed as an alternative to DDT, this organochloride is an extremely persistent pollutant that is biomagnified (like DDT) as it passes through the food chain

ectoparasite—a parasite that lives on the outside of its host, such as fleas and ticks

electroshocking—using an electric generator to emit a nonlethal current through water to temporarily stun fish so they can be sampled

Emiquon—a National Wildlife Refuge and a Nature Conservancy preserve, both located on the Illinois River, near Havana, Illinois

endemic—an organism that is native or restricted to a certain definable area

entomophily—pollination by insects

Entrada Sandstone—a rock formation deposited during the Jurassic period between 180 and 140 million years ago in various environments; tidal mudflats, beaches, and sand dunes

Ephemeroptera—an order on insects called mayflies; the nymphs are aquatic, and the winged adults live for only a short period

ephydrid/Ephydridae—a family of tiny flies that inhabit shorelines; some species can live in crude petroleum, while others prefer briny waters

epiphytic—a plant that grows on another plant and uses it for support but not nutrients

erythrism—refers to an unusual reddish color in an animal's fur, hair, skin, or exoskeleton

exotic invasive species—a plant, fungus, or animal species not native to a specific location (introduced species) that spreads to a degree that causes damage to the environment

exudate—a fluid emitted by an organism, often through pores or a wound

fecal/feces—a solid waste product from the digestive tract of an animal

feeding bed—a beaver activity whereby the animals store sticks and logs in a pile in a pond, subse-quently eating the underbark, often during cold weather

fluvial geomorphology—the study of the physical processes associated with streams and rivers

foci—the center of interest of a given entity

food chain or food web—a network (linear or weblike) of links showing the relationships of organisms by the food they eat

Georgian—an architectural style prevalent between 1720 and 1830, used on many contemporary university campuses

grease ice—a thin layer of soupy ice crystals that makes the water's surface resemble an oil slick

guano—the excrement of many species of communal nesting birds (and bats)

Hemiptera—the largest order of insects that undergo incomplete metamorphosis; includes true bugs, cicadas, leafhoppers, aphids, and so on

hermaphrodite—an organism that has both male and female sex organs

herptile—a term applied to reptiles and amphibians

Hesperiidae—a butterfly characterized by quick darting flight and antennae with a sickle-shaped tip

hill prairie—island-like patches of prairie vegetation growing on otherwise wooded steep slopes that face south or southwest

Hippoboscidae—a group of flies that are obligate parasites of mammals and birds

hogback—a long, narrow ridge with steep sides

hognose snake—a colubrid snake endemic to North America that uses "playing dead" as a defensive strategy; they are mildly venomous, but seldom bite humans

Holocene—the second epoch of the Quaternary Period, following the Pleistocene; the official current epoch (as opposed to the Anthropocene)

hoodoo—a tall, thin spire of soft rock capped by harder rock that protrudes from the floor of a drainage basin or a badland, caused by differential erosion

horseshoe crab—a marine Arthropod, related to Arachnids, in the order Xiphosurida

Huygens principle—when every point on a wave front

acts as a source of secondary waves of light that spread out in all directions

hybrid—the offspring of two plants or animals that are of different species

ice spiders—flow patterns in snow covering ice created by water flowing from under the ice into the snow

Illinois Steward—a nature magazine produced by University of Illinois Extension

incomplete metamorphosis—gradual insect development where the young resemble the adults, except for size; there is no pupal stage

in flagrante delicto—caught in the act; usually applies to sexual activity

interspecific hybridization—the results of mating between different species

introgressive hybridization or introgression—the transfer of genetic information between a hybrid and a member of either of the hybrid's parent species, that is, backcrossing

intromittent organ—an external organ from a male animal that is used to deliver sperm to a female during copulation

in situ—refers to in the original place

instar—the developmental stage in insects between molts

Jolliet and Marquette—early explorers of the Illinois River valley and other parts of North America

Jurassic Period—a geological time period from 200 to 145 millions years ago; known as the Age of Reptiles

kairomone—a chemical emitted by one organism that can be detected by another and used for its advantage (for example, a parasite finds its host by its smell)

leishmaniasis—a disease caused by a protozoan parasite and spread by the bite of various species of sandflies; symptoms are skin sores that will not heal

lek or lekking—a gathering of male animals that engage in competitive displays to attract females

Lepidoptera—the order of insects that includes butterflies and moths; wings of Lepidoptera are covered with scales

lexicon—the words associated with a particular language

lick—a place where animals go to obtain essential nutrients from a mineral deposit

light pollution—excessive, misdirected, or obtrusive artificial light that makes it difficult to see stars in the night sky

light spectrum—that part of the electromagnetic spectrum that is visible to the human eye; also called visible light

lithification—the process whereby sediments are compacted under pressure and turn to stone

living fossil—a species living today that is similar to a species known only from fossils, such as horseshoe crab, coelocanth fish, or gingko tree

macroinvertebrate—a macroscopic animal without a backbone, such as crayfish, stonefly, mayfly, and others

mangrove—salt-tolerant trees that are adapted to live in coastal conditions

mariposa—the Spanish word for butterfly

Megaloptera—an order of insects that contains alderflies, dobsonflies, and fishflies; the name derives from their large, clumsy wings

melittophily—pollination of flowers by bees

Mendelian genetics—inheritance that follows the laws discovered by Gregor Mendel in 1865-66

mesophytic—needing only a moderate amount of water; often refers to a plant

metamorphose—to change into a completely different form or appearance

methyl eugenol—an essential oil, the methyl ether of eugenol, that is an attractant for fruit flies

midden—a refuse heap

mimic or mimicry—the action or art of imitating something, often to gain benefit from the similarity

mineralization—the process of an organic substance becoming impregnated with inorganic materials

mulga—any of a number of *Acacia* species of plants

Müllerian mimicry—where two or more distasteful species evolve to mimic each other's warning signals or colors; predators have fewer patterns to learn to avoid

Neuroptera—an order of insects, including lacewings, mantisflies, and ant lions, that have netlike wings as adults

nictitating membrane—a translucent or transparent third eyelid that can be drawn across the eye in animals for protection

obligate parasite—an organism that cannot complete its life cycle without exploiting a suitable host

Odonata—an order of insects that contain the dragonflies and damselflies

Okavango Delta—a large inland delta in Botswana where the Okavango River flows into the Kalahari Desert; the water eventually evaporates

ommatidia—the optical units that make up the compound eye of an insect

organochlorine—any of a large group of pesticides or other synthetic chemicals with aromatic molecules that contain chlorine

outback—refers to the vast, remote, and arid interior of Australia

overhang or undercut—a rock formation that extends outward, forming a shelter cave or shelter bluff

oxic—having oxygen present or involved in a process

pancake ice—round pieces of ice that have a raised edge

Pantanal—a natural region of Brazil, Bolivia, and Paraguay that encompasses the largest tropical wetland on Earth

parasite/parasitic—an organism that lives on or in another organism (called the host) and derives benefits at the expense of the host

parthenogensis—reproduction from an egg that occurs without fertilization

pelagic—relating or referring to the open sea

Pennsylvanian—a geological subperiod of the Carboniferous Period

permafrost—a thick subsurface layer of soil that remains frozen throughout the year; mostly found in polar regions

Petoskey stone—a pebble-shaped rock composed of fossilized rugose coral

phalaenophily—pollination of plants by moths

pheromone—a chemical released into the environment by a mammal or insect that affects the behavior of others of that species; pheromones are often sex attractants

photoperiod—day length, or the period of time each day that an organism receives light

phototactic—movement of an organism to or away from a light source

Phragmites—called "common reed," a large perennial grass found in wetlands that often takes over

planidia—an insect larva that, in its first stage, is highly active and later molts into a legless parasite

plastron—the lower nearly flat shell of a turtle; belly or ventral surface

Plecoptera—an order of insects (stoneflies) whose nymphs are aquatic and intolerant of pollution

pointillism—a painting technique where small distinct dots of color form the image

polarizing filter—a piece of optical glass placed in front of a camera lens to darken skies and manage reflections from bright surfaces

pollinia—a mass of pollen grains produced by an anther of the flower, usually orchids

protozoology—the scientific study of one-celled organisms

provenance—the documented history of an object or specimen

primary consumer—an animal that feeds on producers (green plants); may also be called herbivore

primary producer—organisms in an ecosystem that produce biomass from inorganic materials (for example, plants from photosynthesis)

psychophily—pollination of plants by butterflies

pterodactyl/pteranodon—flying reptiles prevalent in the late Triassic Period (also called pterosaurs)

puddleclub—a gathering of bachelor male butterflies at a moist spot to imbibe nutrients and also to attract female butterflies

raptor—in contemporary usage, a bird of prey (hawk, eagle, and so on)

raptorial—a limb or other organ adapted for grabbing prey

Reduviidae—a large family of terrestrial ambush predators in the insect order Hemiptera

refractive index—a number that describes how light

propagates through a medium

riparian—refers to vegetation or habitats found along the banks of a stream

rookery—a colony of breeding animals, usually birds

run or riffle—shallow areas in streams where water runs faster or is agitated by rocks

sandstone—sedimentary rock composed mostly of sand-size minerals or grains of rock

San Rafael Swell—a large geological feature in south-central Utah; a giant dome-shaped anticline of sandstone, shale, and limestone

seasonal affective disorder—a type of depression that corresponds to seasonal changes in light

scent mound—a pile of mud and debris, smeared with smelly castoreum, created by a beaver around the perimeter of its territory to keep out other beavers

scute—a thick, bony plate on a turtle's shell; the back of a crocodile or a dinosaur (for example, *Stegosaurus*)

senesce—to deteriorate with age; to grow old

sexual selection—when members of one sex choose mates of the other sex with whom to mate; often forces members of one sex to compete with each other for the right to mate

shale—sedimentary rock composed primarily of mud; a mix of flakes of clay minerals and silt

sheep chill index—windchill that is reported in sheep-raising areas to help minimize stress on sheep during shearing or transportation

shelter bluff or cave—a shallow cave-like opening at the base of a cliff, formed when resistant cap rocks are underlain by softer, easily eroded materials

siltstone—a sedimentary rock formed primarily from silt; finer than sandstone, but coarser than claystone

skipper—members of the butterfly family Hesperiidae; these butterflies have mothlike bodies, hooked antennal tips, and rapid flight

Snell's law—a formula used to describe light waves passing through the boundary between two different media (for example, air and water, air and ice)

soybean aphid—an introduced insect (*Aphis glycines*) from China that is a pest of U.S. soybeans

sperm precedence—the tendency of a female who has bred with several males to give birth to their offspring in unequal proportions; the last male to breed is the sperm used to fertilize the female's eggs

spillway—a structure that provides the controlled release of water from a dam or levee

Spiraea—a genus of hardy deciduous-leaved shrubs and herbs

state champion (tree)—the largest living example of a particular species of tree in a state

stonefly—members of the insect order Plecoptera

surface tension—at liquid/air junctions, the result of the greater attraction of liquid molecules to each other than to the molecules of air

sward—a large expanse of short grass

taxonomic revision—a reordering of various groups that have been classified in a certain manner; revisions are often undertaken when new information reveals the present classification is incorrect

Testudinae—the order of reptiles that comprises the turtles, terrapins, and tortoises

Title IX—a civil rights law that prevents discrimination on the basis of sex in any educational program or activity that receives federal funding

trilobite—a fossil group of extinct marine arthropods

trophic structure—the relationship of organisms to each other in the context of a food web

trypanosome—a single-celled protozoan parasite with a long flagellum that infests blood

tufa—a porous rock of calcium carbonate formed by precipitation from water

Yucatán—a large peninsula in southeastern Mexico that separates the Caribbean Sea from the Gulf of Mexico

REFERENCES

Ackerman, Diane. *The Moon by Whale Light and Other Adventures among Bats, Penguins, Crocodilians, and Whales.* New York: Vintage Books, 1991.

Audubon, John J., and Maria R. Audubon. *Audubon and His Journals: With Zoölogical and Other Notes by Elliott Coues.* Edited by Maria R. Audubon and Elliot Coues. Vol. 1. New York: Charles Scribner's Sons, 1897.

Bouseman, John K., James G. Sternburg, and James R. Wiker. "A Field Guide to the Skipper Butterflies of Illinois." *Illinois Natural History Survey Manual* 11 (2010).

Braby, Michael F. *Butterflies of Australia: Their Identification, Biology and Distribution.* Vol. 1. Australia: CSIRO, 2011.

Bradbury, Ray. *Dandelion Wind.* New York: Doubleday, 1957.

Brown, L. H., E. K. Urban, and K. Newman. *The Birds of Africa.* Vol. 1. Princeton, N.J.: Princeton University Press, 1982.

Carson, Rachel. *Silent Spring.* Cambridge, Mass.: Riverside Press, 1962.

Case, Frederick W., Jr., and Roberta Case. *Trilliums.* Portland, Ore.: Timber Press, 1997.

Conniff, Richard. *House of Lost Worlds.* New Haven, Conn.: Yale University Press, 2016.

Copeland, Kathy, and Craig Copeland. *Hiking from Here to Wow: Utah Canyon Country.* Berkeley, Calif.: Wilderness Press, 2010.

Dickson, James G., ed. *The Wild Turkey Biology and Management.* A National Wild Turkey Federation and USDA Forest Service Book. Mechanicsburg, Pa.: Stackpole Books, 1992.

Dodd, Kenneth C., Jr. *The Amphibians of Great Smoky Mountains National Park.* Knoxville: University of Tennessee Press, 2004.

Dunne, Pete. *Pete Dunne's Essential Field Guide Companion.* Boston: Houghton Mifflin, 2006.

Eaton, Eric R., and Kenn Kaufmann. *Kaufmann Field Guide to Insects of North America.* Boston: Houghton Mifflin, 2007.

Eiseley, Loren C. *The Immense Journey.* New York: Vintage Books, 1959.

Emmons, Louise. *Neotropical Rainforest Mammals: A Field Guide.* 2nd ed. Chicago: University of Chicago Press, 1997.

Estes, Richard D. *The Safari Companion: A Guide to Watching African Mammals.* White River Junction, Vt.: Chelsea Green, 1999.

Ferris, Paul. *Dylan Thomas: The Biography.* Washington, D.C.: Counterpoint, 2000.

Forbes, S. A. *Bulletin of the Illinois State Laboratory of Natural History* 1 (1876).

Frison, Theodore H. "The Stoneflies or Plecoptera of Illinois." *Illinois Natural History Survey Bulletin* 20 (1935): 281–471.

Fry, C. H., and S. Keith, eds. *The Birds of Africa.* Vol. 7. Princeton, N.J.: Princeton University Press, 2000.

Fry, C. H., S. Keith, and E. K. Urban, eds. *The Birds of Africa.* Vol. 3. Princeton, N.J.: Princeton University Press, 1988.

Garga, N. "Aposematism in Water Mites (Acari: Hydracarina): An Anti-predator Defense Mechanism, a Phylogenetic Hold-over and Protection from Damaging Light." M.Sc. thesis, Queen's University, 1996.

Great Smoky Mountains Natural History Association. *Hiking Trails of the Smokies.* Gatlinburg, Tenn.: Great Smoky Mountains Natural History Association, 1994.

Grove, Casey. "Grizzly Attacks Teens in Wilderness School." *Alaska Dispatch News* (Anchorage), July 24, 2011.

Holloway, Tabitha. "Report: Coloration in Katydids." In *2012 Proceedings for the Invertebrates in Education and Conservation Conference.* N.p., n.d.

Illinois-Indiana Sea Grant. *Asian Carp: Bighead and Silver.* Invasive species watch card. IL-IN Sea Grant and Illinois Natural History Survey, 2011.

Imperato, Ferrante. *Historia naturale di Ferrante Imperato napolitano: Nella quale ordinatamente si tratta della diversa condition di minere, pietre pretiose, & altre curiosità; Con varie historie di piante, & animali, sin'hora non date in luce.* Venetia: Presso Combi and La Noù, 1672.

Jackson, Marion T., ed. *The Natural Heritage of Indiana.* Indianapolis: Indiana Academy of Sciences, 1997.

Jagt, Kerry van der. "Cool Birds and Punk Rockers."

March 13, 2011. http://www.traveller.com.au.

Jeffords, M. R. "Capturing Mystery." *Illinois Steward* 12, no. 1 (2003).

——. "Illinois from Above." *Illinois Steward* 13, no. 1 (2004).

——. "Just the Slime." *Illinois Steward* 11, no. 4 (2003).

——. "Monarch Magic." *Illinois Steward* 16, no. 3 (2007).

——. "On Extinction." *Illinois Steward* 9, no. 4 (2001).

——. "When Pictures Aren't Enough." *Illinois Steward* 6, no. 4 (1998).

——. "Why, Why, Why, Take Pictures?" *Illinois Steward* 9, no. 2 (2000).

Jeffords, M. R., S. L. Post, and J. R. Wiker. *Butterflies of Illinois: A Field Guide*. Manual 14. Champaign: Illinois Natural History Survey, 2014.

Jeffords, Michael, and Susan Post. *Exploring Nature in Illinois*. Urbana: University of Illinois Press, 2014.

Jeffords, M. R., and D. W. Webb. "Mating Differences in the Madicolous Crane Fly *Dactylolabis montana* (Osten Sacken) (Diptera: Tipluidae)." *Journal of the Kansas Entomological Society* 75, no. 2 (2002): 138–40.

Kirwan, Guy M., and Graeme Green. *Cotingas and Manakins*. Princeton, N.J.: Princeton University Press, 2011.

Kricher, John. *A Neotropical Companion: An Introduction to the Animals, Plants, and Ecosystems of the New World Tropics*. Princeton, N.J.: Princeton University Press, 1999.

Langlois, Thomas H., and Marina H. Langlois. "Notes on the Life-History of the Hackberry Butterfly, *Asterocampa celtis* (Bdvl. & Lec.) on South Bass Island, Lake Erie (Lepidoptera: Nymphalidae)." *Ohio Journal of Science* 64 (1964): 1–11.

Leopold, Aldo. *A Sand County Almanac and Sketches Here and There*. Oxford: Oxford University Press, 1949.

Lockwood, Mark W., William B. McKinney, James N. Patton, and Barry R. Zimmer. *ABA Bird-Finding Guide: A Birder's Guide to the Rio Grande Valley*. Colorado Springs: American Birding Association, 1999.

Lorenz, Konrad. *Studies in Animal and Human Behavior*. Cambridge, Mass.: Harvard University Press, 1971.

Lovich, Jeff. "Aggressive Basking Behavior in Eastern Painted Turtles (*Chrysemys picta picta*)." *Herpetologica* 44, no. 2 (1988): 197–202.

Lyons, Stephen J. *A View from the Inland Northwest*. Guilford, Conn.: Globe Pequot Press, 2004.

Mohlenbrock, Robert. H. *Vascular Flora of Illinois: A Field Guide*. Carbondale: Southern Illinois University Press, 2014.

Mono Lake Story. N.d. http://www.monolake.org/about/story.

Müller-Schwarze, Dietland, and Lixing Sun. *The Beaver: Natural History of a Wetland Engineer*. Ithaca, N.Y.: Cornell University Press, 2003.

Murphy, Robert Cushman. *Logbook for Grace*. New York: Macmillan, 1947.

Neruda, Pablo. *Art of Birds*. Austin: University of Texas Press, 1985.

Oberhauser, Karen, and Michael A. Quinn. *Milkweed, Monarchs and More: A Field Guide to the Invertebrate Community in the Milkweed Batch*. Glenshaw, Pa.: Bas Relief, 2003.

Paz, Alberto, and Valorie Hart. *Gotta Tango*. Champaign, Ill.: Human Kinetics, 2008.

Pegg, Mark. "Asian Carp." Sidebar in "Exotic, Invasive Species in Illinois," by Robert N. Wiedenmann. *Illinois Steward* (Spring 2002): 11–18.

Post, Susan. "Demons of the Dust." *Illinois Steward* 13, no. 4 (2004): 2-4.

——. "Spiny Softshell Turtle." *Illinois Steward* 14, no. 1 (2005): 2–3.

——. "The Tree Squirrel Bot Fly." *Illinois Natural History Survey Reports*, no. 406 (Summer 2011).

Post, Susan L., and Michael Jeffords. "Treasures of the Cache." *Illinois Steward* 15, no. 1 (2006).

Rafferty, John P. "Anthropocene Epoch." 2016. http://www.britannica.com/science/Anthropocene-Epoch.

Reber, Robert J., and Michael Jeffords. "A Leopold Experience." *Illinois Steward* 11, no. 2 (2002).

——. "Old Friends." *Illinois Steward* 7, no. 2 (1998).

Richardson, W. Mark. "A Skeptic's View of Wonder." *Scientific American* (1992).

Ridgely, Robert S., and Paul J. Greenfield. The *Birds of Ecuador: A Field Guide*. Ithaca, N.Y.: Cornell University Press, 2001.

Sedman, Y., and D. F. Hess. "The Butterflies of West Central Illinois." *Western Illinois University Series in Biological Sciences* 11 (1985): 1–117.

Smith, James Edward, Sir, 1759–1828. *The Natural History of the Rarer Lepidopterous Insects of Georgia: Including Their Systematic Characters, the Particulars of Their Several Metamorphoses, and the Plants on Which They Feed.* London, 1797.

Stannard, L. J. "The Distribution of Periodical Cicadas in Illinois." *Illinois Natural History Survey Biological Notes* 91 (1975): 3–12.

Steinbeck, John. *East of Eden.* New York: Penguin Books, 2002.

Stokstad, Erik. "Science Shot: First Example of Tool Use in Reptiles." 2013. http://www.sciencemag. org/news/2013/12/scienceshot-first-example-tool-use-reptiles.

Teale, Edwin Way. *North with the Spring.* New York: St. Martin's Press, 1951.

Tinbergen, Niko. *The Herring Gull's World.* London: Collins, 1953.

Zaborski, E. R., and L. A. Soeken Gittenger. "*Amynthas hupeiensis* (Michaelsen, 1895) (Oligochaeta: Megascolecidae) in Illinois, USA, with Observations on Worm Circling." *Megadrilogica* 8, no. 4 (2001): 13–16.

Zim, Herbert H., and Alexander C. Martin. *Trees.* Golden Nature Guide. New York: Golden Press, 1956.

INDEX

MICHAEL R. JEFFORDS is a retired entomologist from the University of Illinois Prairie Research Institute, Illinois Natural History Survey, where he served as a research scientist and the education/outreach coordinator. Jeffords is a freelance writer and photographer, and has authored or edited four books, including *Exploring Nature in Illinois* and *A Field Guide to Illinois Butterflies* (both with Susan Post). He was also staff photographer for the Illinois Steward magazine for nearly 20 years. While he and his wife (Susan Post) travel widely across the U.S. and the world, his home base continues to be Champaign, IL.

SUSAN L. POST is a retired research scientist from the University of Illinois Prairie Research Institute, Illinois Natural History Survey and a freelance writer and photographer. Susan was the staff writer for the Illinois Steward magazine and has authored six books, including two editions of *Hiking Illinois*. Sue's passions are travel, birding, photography, and small mammals (especially guinea pigs). Susan and Michael have been married for 34 years, and have collaborated on many joint projects on education, issues of biodiversity, and in the art and science of communicating nature to a wide audience.

The University of Illinois Press
is a founding member of the
Association of American University Presses.

Designed by Danielle Ruffatto
Composed in 08/12 PT Sans
with Trend RH Slab and Trend RH Sans display
at the University of Illinois Press
Manufactured by Four Color Print Group

University of Illinois Press
1325 South Oak Street
Champaign, IL 61820-6903
www.press.uillinois.edu